ID0847980

Wojbor A. Woyczyński

A First Course in Statistics for Signal Analysis

Second Edition

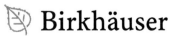 Birkhäuser

Wojbor A. Woyczyński
Department of Statistics
  and Center for Stochastic and Chaotic Processes
  in Sciences and Technology
Case Western Reserve University
10900 Euclid Avenue
Cleveland, OH 44106
USA
waw@case.edu
http://stat.case.edu/~Wojbor/

ISBN 978-0-8176-8100-5   e-ISBN 978-0-8176-8101-2
DOI 10.1007/978-0-8176-8101-2
Springer New York Dordrecht Heidelberg London

Library of Congress Control Number: 2010937639

Mathematics Subject Classification (2010): 60-01, 60G10, 60G12, 60G15, 60G35, 62-01, 62M10,
                                    62M15, 62M20

©Springer Science+Business Media, LLC 2011
All rights reserved. This work may not be translated or copied in whole or in part without the written
permission of the publisher (Springer Science+Business Media, LLC, 233 Spring Street, New York,
NY 10013, USA), except for brief excerpts in connection with reviews or scholarly analysis. Use in
connection with any form of information storage and retrieval, electronic adaptation, computer software,
or by similar or dissimilar methodology now known or hereafter developed is forbidden.
The use in this publication of trade names, trademarks, service marks, and similar terms, even if they are
not identified as such, is not to be taken as an expression of opinion as to whether or not they are subject
to proprietary rights.

Printed on acid-free paper

www.birkhauser-science.com

This book is dedicated to my children:
Martin Wojbor, Gregory Holbrook,
and Lauren Pike.
  They make it all worth it.

Contents

Foreword to the Second Edition

The basic structure of the Second Edition remains the same, but many changes have been introduced, responding to several years' worth of comments from students and other users of the First Edition. Most of the figures have been redrawn to better show the scale of the quantities represented in them, some notation and terminology have been adjusted to better reflect the concepts under discussion, and several sections have been considerably expanded by the addition of new examples, illustrations, and commentary. Thus the original conciseness has been somewhat softened. A typical example here would be the addition of Remark 4.1.2, which explains how one can see the Bernoulli white noise in continuous time as a scaling limit of switching signals with exponential interswitching times. There are also new, more applied exercises as well, such as Problem 9.7.9 on simulating signals produced by spectra generated by incandescent and luminescent lamps.

Still the book remains more mathematical than many other signal processing books. So, at Case Western Reserve University, the course (required for most electrical engineering and some biomedical engineering juniors/seniors) based on this book runs in parallel with a signal processing course that is entirely devoted to practical applications and software implementation. This one–two-punch approach has been working well, and the engineers seem to appreciate the fact that all probability/statistics/Fourier analysis foundations are developed within the book; adding extra mathematical courses to a tight undergraduate engineering curriculum is almost impossible. A gaggle of graduate students in applied mathematics, statistics, and assorted engineering areas also regularly enrolls. They are often asked to make in-class presentations of special topics included in the book but not required of the general undergraduate audience.

Finally, by popular demand, there is now a large appendix which contains solutions of selected problems from each of the nine chapters. Here, most of the credit goes to my former graduate students who served as TAs for my courses: Aleksandra Piryatinska (now at San Francisco State University), Sreenivas Konda (now at Temple University), Dexter Cahoy (now at Louisiana Tech), and Peipei Shi (now at Eli Lilly, Inc.). In preparing the Second Edition, the author took into account useful comments that appeared in several reviews of the original book; the review published in September 2009 in the *Journal of the American Statistical Association* by Charles Boncelet was particularly thorough and insightful.

Cleveland   *Wojbor A. Woyczyński*
May 2010   *http://stat.case.edu/~Wojbor*

Introduction

This book was designed as a text for a first, one-semester course in statistical signal analysis for students in engineering and physical sciences. It had been developed over the last few years as lecture notes used by the author in classes mainly populated by electrical, systems, computer, and biomedical engineering juniors/seniors, and graduate students in sciences and engineering who have not been previously exposed to this material. It was also used for industrial audiences as educational and training materials, and for an introductory time-series analysis class.

The only prerequisite for this course is a basic two- to three-semester calculus sequence; no probability or statistics background is assumed except the usual high school elementary introduction. The emphasis is on a crisp and concise, but fairly rigorous, presentation of fundamental concepts in the statistical theory of stationary random signals and relationships between them. The author's goal was to write a compact but readable book of less than 200 pages, countering the recent trend toward fatter and fatter textbooks.

Since Fourier series and transforms are of fundamental importance in random signal analysis and processing, this material is developed from scratch in Chap. 2, emphasizing the time-domain vs. frequency-domain duality. Our experience showed that although harmonic analysis is normally included in the calculus syllabi, students' practical understanding of its concepts is often hazy. Chapter 3 introduces basic concepts of probability theory, law of large numbers and the stability of fluctuations law, and statistical parametric inference procedures based on the latter.

In Chap. 4 the fundamental concept of a stationary random signal and its autocorrelation structure is introduced. This time-domain analysis is then expanded to the frequency domain by a discussion in Chap. 5 of power spectra of stationary signals. How stationary signals are affected by their transmission through linear systems is the subject of Chap. 6. This transmission analysis permits a preliminary study of the issues of designing filters with the optimal signal-to-noise ratio; this is done in Chap. 7. Chapter 8 concentrates on Gaussian signals where the autocorrelation structure completely determines all the statistical properties of the signal. The text concludes, in Chap. 9, with the description of algorithms for computer simulations of stationary random signals with a given power spectrum density. The routines are based on the general spectral representation theorem for such signals, which is also derived in this chapter.

The book is essentially self-contained, assuming the indispensable calculus background mentioned above. A complementary bibliography, for readers who would like to pursue the study of random signals in greater depth, is described at the end of this volume.

Some general advice to students using this book: The material is deliberately written in a compact, economical style. To achieve the understanding needed for independent solving of the problems listed at the end of each chapter in the Problems and Exercises sections, it is not sufficient to read through the text in the manner you would read through a newspaper or a novel. It is necessary to look at every single statement with a "magnifying glass" and to decode it in your own technical language so that you can use it *operationally* and not just be able to talk about it. The only practical way to accomplish this goal is to go through each section with pencil and paper, explicitly completing, if necessary, routine analytic intermediate steps that were omitted in the exposition for the sake of the clarity of the presentation of the bigger picture. It is the latter that the author wants you to keep at the end of the day; there is no danger in forgetting all the little details if you know that you can recover them by yourself when you need them.

Finally, the author would like to thank Profs. Mike Branicky and Ken Loparo of the Department of Electrical and Computer Engineering, and Prof. Robert Edwards of the Department of Chemical Engineering of Case Western Reserve University for their kind interest and help in the development of this course and comments on the original version of this book. My graduate students, Alexey Usoltsev and Alexandra Piryatinska, also contributed to the editing process, and I appreciate the time they spent on this task. Partial support for this writing project from the Columbus Instruments International Corporation of Columbus, Ohio, Dr. Jan Czekajewski, President, is also gratefully acknowledged.

Four anonymous referees spent considerable time and effort trying to improve the original manuscript. Their comments are appreciated and, almost without exception, their sage advice was incorporated in the final version of the book. I thank them for their help.

Notation

To be used only as a guide and not as a set of formal definitions.

$\mathbf{AV}_x$	time average of signal $x(t)$
BW_n	equivalent-noise bandwidth of the system
$BW_{1/2}$	half-power bandwidth of the system
$\mathbf{C}$	the set of all complex numbers
$\text{Cov}(X, Y) =$	
$\mathbf{E}[(X - \mathbf{E}X)(Y - \mathbf{E}Y)]$	covariance of X and Y
δ_{mn}	Kronecker's delta, $= 0$ if $m \neq n$, and $= 1$ if $m = n$
$\delta(x)$	Dirac delta "function"
$\mathbf{EN}_x$	energy of signal $x(t)$
$\mathbf{E}(X)$	expected value (mean) of a random quantity X
$F_X(x)$	cumulative distribution function (c.d.f.) of a random quantity X
$f_X(x)$	probability density function (p.d.f.) of a random quantity X
$\gamma_X(\tau) =$	
$\mathbf{E}(X(t) - \mu_X)(X(t+\tau) - \mu_X)$	autocovariance function of a stationary signal $X(t)$
$h(t)$	impulse response function of a linear system
$H(f)$	transfer function of a linear system, Fourier transform of $h(t)$
$\|H(f)\|^2$	power transfer function of a linear system
$L_0^2(\mathbf{P})$	space of all zero-mean random quantities with finite variance
$m^\alpha(X) = \mathbf{E}\|X\|^\alpha$	αth absolute moment of a random quantity X
$\mu^k(X) = \mathbf{E}(X^k)$	kth moment of a random quantity X
$N(\mu, \sigma^2)$	Gaussian (normal) probability distribution with mean μ and variance σ^2
P	period of a periodic signal
$\mathbf{P}(A)$	probability of event A
$\mathbf{PW}_x$	power of signal $x(t)$

$Q_X(\alpha) = F_X^{-1}(\alpha)$	α's quantile of random quantity X
R	resolution
$\mathbf{R}$	the set of all real numbers
$\rho_{X,Y} = \mathrm{Cov}(X,Y)/(\sigma_X \sigma_Y)$	correlation coefficient of X and Y
$\mathrm{Std}(X) = \sigma_X = \sqrt{\mathrm{Var}(X)}$	the standard deviation of a random quantity X
$S_X(f)$	power spectral density of a stationary signal $X(t)$
$\mathcal{S}_X(f)$	cumulative power spectrum of a stationary signal $X(t)$
T	sampling period
$u(t)$	Heaviside unit step function, $u(t) = 0$, for $t < 0$, and $= 1$, for $t \geq 0$
$\mathrm{Var}(X) = \mathbf{E}(X - \mathbf{E}X)^2 = \mathbf{E}X^2 - (\mathbf{E}X)^2$	the variance of a random quantity X
$W(n)$	discrete-time white noise
$\mathcal{W}(n)$	cumulative discrete-time white noise
$W(t)$	continuous-time white noise
$\mathcal{W}(t)$	the Wiener process
$x(t), y(t),$ etc.	deterministic signals
$X = (X_1, X_2, \ldots, X_d)$	a random vector in dimension d
$x(t) * y(t)$	convolution of signals $x(t)$ and $y(t)$
$X(f), Y(f)$	Fourier transforms of signals $x(t)$ and $y(t)$, respectively
X, Y, Z	random quantities (random variables)
z^*	complex conjugate of complex number z; i.e., if $z = \alpha + j\beta$, then $z^* = \alpha - j\beta$
$\lfloor a \rfloor$	"floor" function, the largest integer not exceeding number a
$\langle . , . \rangle$	inner (dot, scalar) product of vectors or signals
$\Leftrightarrow$	if and only if
$:=$	is defined as

Chapter 1
Description of Signals

Signals are everywhere. Literally. The universe is bathed in the background radiation, the remnant of the original Big Bang, and as your eyes scan this page, a signal is being transmitted to your brain, where different sets of neurons analyze and process it. All human activities are based on the processing and analysis of sensory signals, but the goal of this book is somewhat narrower. The signals we will be mainly interested in can be described as *data* resulting from quantitative measurements of some physical phenomena, and our emphasis will be on data that display *randomness* that may be due to different causes, be it errors of measurements, the algorithmic complexity, or the chaotic behavior of the underlying physical system itself.

1.1 Types of Random Signals

For the purpose of this book, signals will be functions of the real variable t interpreted as time. To describe and analyze signals, we will adopt the functional notation: $x(t)$ will denote the value of a nonrandom signal at time t. The values themselves can be real or complex numbers, in which case we will symbolically write $x(t) \in \mathbf{R}$ or, respectively, $x(t) \in \mathbf{C}$. In certain situations it is necessary to consider vector-valued signals with $x(t) \in \mathbf{R}^d$, where d stands for the dimension of the vector $x(t)$ with d real components.

Signals can be classified into different categories depending on their features. For example:

- *Analog signals* are functions of continuous time, and their values form a continuum. *Digital signals* are functions of discrete time dictated by the computer's clock, and their values are also discrete and dictated by the resolution of the system. Of course, one can also encounter mixed-type signals which are sampled at discrete times but whose values are not restricted to any discrete set of numbers.
- *Periodic signals* are functions whose values are periodically repeated. In other words, for a certain number $P > 0$, we have $x(t + P) = x(t)$, for any t. The number P is called the *period of the signal*. *Aperiodic signals* are signals that are not periodic.

W.A. Woyczyński, *A First Course in Statistics for Signal Analysis*,
DOI 10.1007/978-0-8176-8101-2_1, © Springer Science+Business Media, LLC 2011

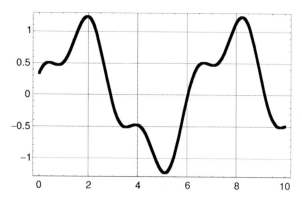

Fig. 1.1.1 The signal $x(t) = \sin(t) + \frac{1}{3}\cos(3t)$ [V] is analog and periodic with period $P = 2\pi$ [s]. It is also deterministic

- *Deterministic signals* are signals not affected by random noise; there is no uncertainty about their values. *Random signals*, often also called *stochastic processes*, include an element of uncertainty; their analysis requires the use of statistical tools, and providing such tools is the principal goal of this book.

For example, the signal $x(t) = \sin(t) + \frac{1}{3}\cos(3t)$ [V] shown in Fig. 1.1.1 is deterministic, analog, and periodic with period $P = 2\pi$ [s]. The same signal, digitally sampled during the first 5 s at time intervals equal to 0.5 s, with resolution 0.01 V, gives tabulated values:

t	0.5	1.0	1.5	2.0	2.5	3.0	3.5	4.0	4.5	5.0
$x(t)$	0.50	0.51	0.93	1.23	0.71	−0.16	0.51	−0.48	−0.78	−1.21

This sampling process is called the *analog-to-digital conversion*: Given the *sampling period* T and the *resolution* R, the digitized signal $x_d(t)$ is of the form

$$x_d(t) = R \left\lfloor \frac{x(t)}{R} \right\rfloor, \quad \text{for} \quad t = T, 2T, \dots, \tag{1.1.1}$$

where the (convenient to introduce here) "floor" function $\lfloor a \rfloor$ is defined as the largest integer not exceeding the real number a. For example, $\lfloor 5.7 \rfloor = 5$, but $\lfloor 5.0 \rfloor = 5$ as well.

Note the role the resolution R plays in the above formula. Take, for example, $R = 0.01$. If the signal $x(t)$ takes all the continuous values between $m = \min_t x(t)$ and $M = \max_t x(t)$, then $x(t)/0.01$ takes all the continuous values between $100 \cdot m$ and $100 \cdot M$, but $\lfloor x(t)/0.01 \rfloor$ takes only integer values between $100 \cdot m$ and $100 \cdot M$. Finally, $0.01 \lfloor x(t)/0.01 \rfloor$ takes as its values only all the discrete numbers between m and M that are 0.01 apart.

The randomness of signals can have different origins, be it the quantum *uncertainty principle, the computational complexity* of algorithms, *chaotic behavior* in dynamical systems, or random fluctuations and errors in the measurement of

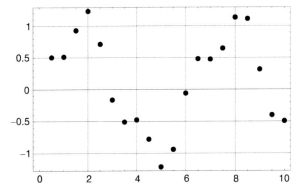

Fig. 1.1.2 The signal $x(t) = \sin(t) + \frac{1}{3}\cos(3t)$ [V] digitally sampled at time intervals equal to 0.5 s with resolution 0.01 V

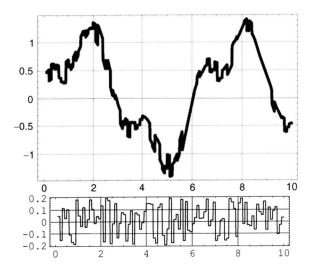

Fig. 1.1.3 The signal $x(t) = \sin(t) + \frac{1}{3}\cos(3t)$ [V] in the presence of additive random noise with a maximum amplitude of 0.2 V. The magnified noise component itself is pictured under the graph of the signal

outcomes of independently repeated experiments.[1] The usual way to study them is via their aggregated statistical properties. The main purpose of this book is to introduce some of the basic mathematical and statistical tools useful in analysis of random signals that are produced under *stationary conditions*, that is, in situations where the measured signal may be stochastic and contain random fluctuations, but

---

[1] See, e.g., M. Denker and W.A. Woyczyński, *Introductory Statistics and Random Phenomena: Uncertainty, Complexity, and Chaotic Behavior in Engineering and Science,* Birkhäuser Boston, Cambridge, MA, 1998.

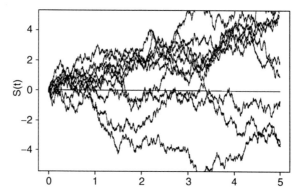

Fig. 1.1.4 Several computer-generated trajectories (sample paths) of a random signal called the *Brownian motion* stochastic process or the *Wiener stochastic process*. Its trajectories, although very rough, are continuous. It is often used as a simple model of *diffusion*. The random mechanism that created different trajectories was the same. Its importance for our subject matter will become clear in Chap. 9

the basic underlying random mechanism producing it does not change over time; think here about outcomes of independently repeated experiments, each consisting of tossing a single coin.

At this point, to help the reader visualize the great variety of random signals appearing in the physical sciences and engineering, it is worthwhile reviewing a gallery of pictures of random signals, both experimental and simulated, presented in Figs. 1.1.4–1.1.8. The captions explain the context in each case.

The signals shown in Figs. 1.1.4 and 1.1.5 are, obviously, not stationary and have a diffusive character. However, their increments (differentials) are stationary and, in Chap. 9, they will play an important role in the construction of the spectral representation of stationary signals themselves. The signal shown in Fig. 1.1.4 can be interpreted as a *trajectory*, or *sample path*, of a *random walker* moving, in discrete-time steps, up or down a certain distance with equal probabilities 1/2 and 1/2. However, in the picture these trajectories are viewed from far away and in accelerated time, so that both time and space appear continuous.

In certain situations the randomness of the signal is due to uncertainty about initial conditions of the underlying phenomenon which otherwise can be described by perfectly deterministic models such as partial differential equations. A sequence of pictures in Fig. 1.1.6 shows the evolution of the system of particles with an initially random (and homogeneous in space) spatial distribution. The particles are then driven by the velocity field $\vec{v}(t, \vec{x}) \in \mathbf{R}^2$ governed by the *2D Burgers equation*

$$\frac{\partial \vec{v}(t, \vec{x})}{\partial t} + \left(\nabla \cdot \vec{v}(t, \vec{x}) \right) \vec{v}(t, \vec{x}) = D \left(\frac{\partial^2 \vec{v}(t, \vec{x})}{\partial x_1} + \frac{\partial^2 \vec{v}(t, \vec{x})}{\partial x_2} \right), \qquad (1.1.2)$$

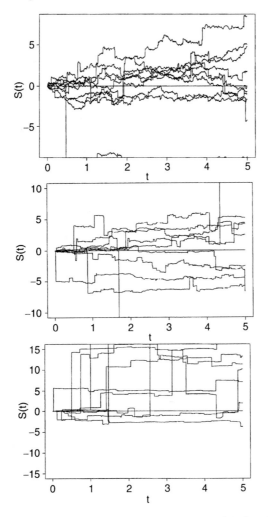

Fig. 1.1.5 Several computer-generated trajectories (sample paths) of random signals called *Lévy stochastic processes* with parameter $\alpha = 1.5, 1$, and 0.75, respectively (from *top* to *bottom*). They are often used to model anomalous diffusion processes wherein diffusing particles are also permitted to change their position by jumping. The parameter α indicates the intensity of jumps of different sizes. The parameter value $\alpha = 2$ corresponds to the Wiener process (shown in Fig. 1.1.4) which has trajectories that have no jumps. In each figure, the random mechanism that created different trajectories was the same. However, different random mechanisms led to trajectories presented in different figures

where $\vec{x} = (x_1, x_2)$, the *nabla* operator $\nabla = \partial/\partial x_1 + \partial/\partial x_2$, and the positive constant D is the coefficient of diffusivity. The initial velocity field is also assumed to be random.

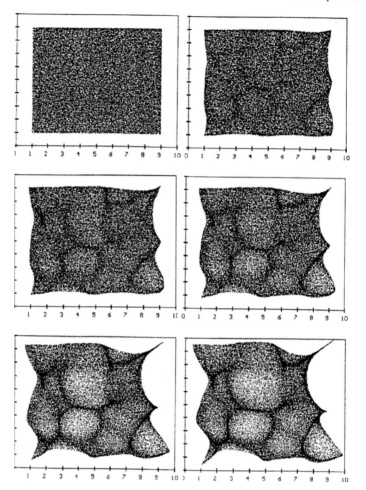

Fig. 1.1.6 Computer simulation of the evolution of passive tracer density in a turbulent velocity field with random initial distribution and random "shot-noise" initial velocity data. The simulation was performed for 100,000 particles. The consecutive frames show the location of passive tracer particles at times $t = 0.0, 0.3, 0.6, 1.0, 2.0, 3.0$ s

1.2 Characteristics of Signals

Several physical characteristics of signals are of primary interest.

- *The time average of the signal:* For analog, continuous-time signals, the time average is defined by the formula

$$\mathbf{AV}_x = \lim_{T \to \infty} \frac{1}{T} \int_0^T x(t)\, dt, \qquad (1.2.1)$$

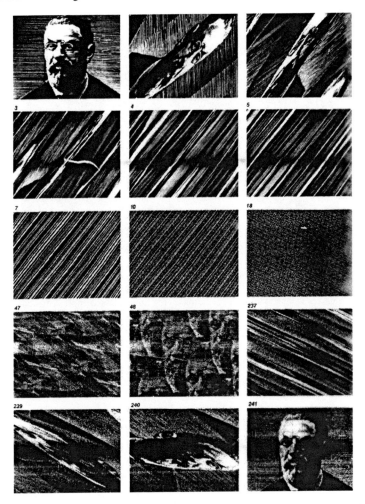

Fig. 1.1.7 Some deterministic signals (in this case, the images) transformed by deterministic systems can appear random. Above is a series of iterated transformations of the original image via a fixed linear 2D mapping (matrix). The number of iterations applied is indicated in the *top left corner* of each image. The curious behavior of iterations – the original image first dissolving into seeming randomness only to return later to an almost original condition – is related to the so-called ergodic behavior. Thus irreverently transformed is Prof. Henri Poincaré (1854–1912) of the University of Paris, the pioneer of ergodic theory of stationary phenomena[2]

and for digital, discrete-time signals which are defined only for the time instants $t = n$, $n = 0, 1, 2, \ldots, N - 1$, it is defined by the formula

$$\mathbf{AV}_x = \frac{1}{N} \sum_{n=0}^{N-1} x(nT). \qquad (1.2.2)$$

---

[2] From *Scientific American*, reproduced with permission. © 1986, James P. Crutchfield.

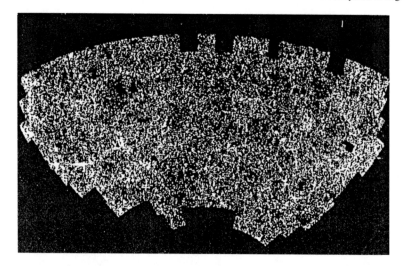

Fig. 1.1.8 A signal (again, an image) representing the large-scale and apparently random distribution of mass in the universe. The data come from the APM galaxy survey and show more than two million galaxies in a section of sky centered on the South Galactic pole. The *adhesion model* of the large-scale mass distribution in the universe uses Burgers' equation to model the relevant velocity fields[3]

For periodic signals, it follows from (1.2.1) that

$$\mathbf{AV}_x = \frac{1}{P} \int_0^P x(t)\, dt, \tag{1.2.3}$$

so that, for the signal $x(t) = \sin t + (1/3)\cos(3t)$ pictured in Fig. 1.1.1, the time average is 0, as both $\sin t$ and $\cos(3t)$ integrate to zero over the period $P = 2\pi$.

- *Energy of the signal:* For an analog signal $x(t)$, the total energy is

$$\mathbf{EN}_x = \int_0^\infty |x(t)|^2\, dt, \tag{1.2.4}$$

and for digital signals it is

$$\mathbf{EN}_x = \sum_{n=0}^\infty |x(nT)|^2 \cdot T. \tag{1.2.5}$$

Observe that the energy of a periodic signal, such as the one from Fig. 1.1.1 is necessarily infinite if considered over the whole positive timeline. Also note that since in what follows it will be convenient to consider complex-valued signals, the above formulas include notation for the square of the modulus of a complex number: $|z|^2 = (\operatorname{Re} z)^2 + (\operatorname{Im} z)^2 = z \cdot z^*$; more about it in the next section.

---

[3] See, e.g., W.A. Woyczyński, *Burgers–KPZ Turbulence – Göttingen Lectures*, Springer-Verlag, New York, 1998.

- *Power of the signal:* Again, for an analog signal, the (average) power is

$$\mathbf{PW}_x = \lim_{T \to \infty} \frac{1}{T} \int_0^T |x(t)|^2 \, dt, \tag{1.2.6}$$

and for a digital signal it is

$$\mathbf{PW}_x = \lim_{N \to \infty} \frac{1}{NT} \sum_{n=0}^{N-1} |x(nT)|^2 \cdot T = \lim_{N \to \infty} \frac{1}{N} \sum_{n=0}^{N-1} |x(nT)|^2. \tag{1.2.7}$$

As a consequence, for an analog periodic signal with period P,

$$\mathbf{PW}_x = \frac{1}{P} \int_0^P |x(t)|^2 \, dt. \tag{1.2.8}$$

For example, for the signal in Fig. 1.1.1,

$$
\begin{aligned}
\mathbf{PW}_x &= \frac{1}{2\pi} \int_0^{2\pi} \left(\sin t + (1/3) \cos(3t) \right)^2 dt \\
&= \frac{1}{2\pi} \int_0^{2\pi} \left(\sin^2 t + \frac{2}{3} \sin t \cos(3t) + \frac{1}{9} \cos^2(3t) \right) dt \\
&= \frac{1}{2\pi} \int_0^{2\pi} \left(\frac{1}{2}(1 - \cos(2t)) + \frac{2}{3}\frac{1}{2}(\sin(4t) - \sin(2t)) + \frac{1}{9}\frac{1}{2}(1 - \cos(6t)) \right) dt \\
&= \frac{1}{2\pi} \left(\frac{1}{2}2\pi + \frac{1}{9}\frac{1}{2}2\pi \right) = \frac{5}{9}. \tag{1.2.9}
\end{aligned}
$$

The above routine calculation, deliberately carried out here in detail, was somewhat tedious because of the need for various trigonometric identities. To simplify such manipulations and make the whole theory more elegant, we will introduce in the next section a complex number representation of the trigonometric functions via the so-called de Moivre formulas.

Remark 1.2.1 (Timeline infinite in both direction). Sometimes it is convenient to consider signals defined for all time instants t, $-\infty < t < +\infty$, rather than just for positive t. In such cases all of the above definitions have to be adjusted in obvious ways, replacing the one-sided integrals and sums by two-sided integrals and sums, and adjusting the averaging constants correspondingly.

1.3 Time-Domain and Frequency-Domain Descriptions of Periodic Signals

The time-domain description. The trigonometric functions

$$x(t) = \cos(2\pi f_0 t) \qquad \text{and} \qquad y(t) = \sin(2\pi f_0 t)$$

represent a harmonically oscillating signal with period $P = 1/f_0$ (measured, say, in seconds [s]), and the frequency f_0 (measured, say, in cycles per second, or Hertz [Hz]), and so do the trigonometric functions

$$x(t) = \cos(2\pi f_0(t + \theta)) \quad \text{and} \quad y(t) = \sin(2\pi f_0(t + \theta))$$

shifted by the phase shift θ. The powers

$$\text{PW}_x = \frac{1}{P} \int_0^P \cos^2(2\pi f_0 t)\, dt = \frac{1}{P} \int_0^P \frac{1}{2}(1 + \cos(4\pi f_0 t))\, dt = \frac{1}{2}, \quad (1.3.1)$$

$$\text{PW}_y = \frac{1}{P} \int_0^P \sin^2(2\pi f_0 t)\, dt = \frac{1}{P} \int_0^P \frac{1}{2}(1 - \cos(4\pi f_0 t))\, dt = \frac{1}{2}, \quad (1.3.2)$$

using the trigonometric formulas from Table 1.3.1. The phase shifts obviously do not change the power of the above harmonic signals.

Taking their linear combination (like the one in Fig. 1.1.1), with amplitudes A and B, respectively,

$$z(t) = Ax(t) + By(t) = A\cos(2\pi f_0(t + \theta)) + B\sin(2\pi f_0(t + \theta)), \quad (1.3.3)$$

also yields a periodic signal with frequency f_0. For a signal written in this form, we no longer need to include the phase shift explicitly since

$$\cos(2\pi f_0(t + \theta)) = \cos(2\pi f_0 t)\cos(2\pi f_0 \theta) - \sin(2\pi f_0 t)\sin(2\pi f_0 \theta)$$

Table 1.3.1 Trigonometric formulas

$$\sin(\alpha \pm \beta) = \sin\alpha\cos\beta \pm \sin\beta\cos\alpha$$

$$\cos(\alpha \pm \beta) = \cos\alpha\cos\beta \mp \sin\alpha\sin\beta$$

$$\sin\alpha + \sin\beta = 2\sin\frac{\alpha + \beta}{2}\cos\frac{\alpha - \beta}{2}$$

$$\sin\alpha - \sin\beta = 2\cos\frac{\alpha + \beta}{2}\sin\frac{\alpha - \beta}{2}$$

$$\cos\alpha + \cos\beta = 2\cos\frac{\alpha + \beta}{2}\cos\frac{\alpha - \beta}{2}$$

$$\cos\alpha - \cos\beta = -2\sin\frac{\alpha + \beta}{2}\sin\frac{\alpha - \beta}{2}$$

$$\sin^2\alpha - \sin^2\beta = \cos^2\beta - \cos^2\alpha = \sin(\alpha + \beta)\sin(\alpha - \beta)$$

$$\cos^2\alpha - \sin^2\beta = \cos^2\beta - \sin^2\alpha = \cos(\alpha + \beta)\cos(\alpha - \beta)$$

$$\sin\alpha\cos\beta = \frac{1}{2}\Big[\sin(\alpha + \beta) + \sin(\alpha - \beta)\Big]$$

$$\cos\alpha\cos\beta = \frac{1}{2}\Big[\cos(\alpha + \beta) + \cos(\alpha - \beta)\Big]$$

$$\sin\alpha\sin\beta = \frac{1}{2}\Big[\cos(\alpha - \beta) - \cos(\alpha + \beta)\Big]$$

and

$$\sin(2\pi f_0(t + \theta)) = \sin(2\pi f_0 t)\cos(2\pi f_0 \theta) + \cos(2\pi f_0 t)\sin(2\pi f_0 \theta),$$

so that

$$z(t) = a\cos(2\pi f_0 t) + b\sin(2\pi f_0 t), \tag{1.3.4}$$

with the new amplitudes

$$a = A\cos(2\pi f_0 \theta) + B\sin(2\pi f_0 \theta) \quad \text{and} \quad b = B\cos(2\pi f_0 \theta) - A\sin(2\pi f_0 \theta).$$

The power of the signal $z(t)$, in view of (1.3.1 and 1.3.2), is given by the Pythagorean-like formula

$$\begin{aligned}
\mathbf{PW}_z &= \frac{1}{P}\int_0^P z^2(t)\,dt = \frac{1}{P}\int_0^P (a\cos(2\pi f_0 t) + b\sin(2\pi f_0 t))^2\,dt \\
&= a^2 \cdot \mathbf{PW}_x + b^2 \cdot \mathbf{PW}_y + 2ab\frac{1}{P}\int_0^P \cos(2\pi f_0 t)\sin(2\pi f_0 t)\,dt \\
&= \frac{1}{2}(a^2 + b^2),
\end{aligned} \tag{1.3.5}$$

because (again, see Table 1.3.1)

$$\frac{1}{P}\int_0^P \cos(2\pi f_0 t)\sin(2\pi f_0 t)\,dt = \frac{1}{P}\int_0^P \frac{1}{2}\sin(4\pi f_0 t)\,dt = 0. \tag{1.3.6}$$

The above property (1.3.6), called *orthogonality* of the sine and cosine signals, will play a fundamental role in this book.

The next observation is that signals

$$z(t) = a\cos(2\pi(mf_0)t) + b\sin(2\pi(mf_0)t), \quad m = 0, 1, 2, \ldots,$$

have frequency equal to the multiplicity m of the *fundamental* frequency f_0, and as such have, in particular, period P (but also period P/m). Their power is also equal to $(a^2 + b^2)/2$. So if we superpose M of them, with possibly different amplitudes a_m and b_m, for different $m = 0, 1, 2, \ldots, M$, the result is a periodic signal

$$\begin{aligned}
x(t) &= \sum_{m=0}^{M}\left(a_m\cos(2\pi(mf_0)t) + b_m\sin(2\pi(mf_0)t)\right) \\
&= a_0 + \sum_{m=1}^{M}\left(a_m\cos(2\pi(mf_0)t) + b_m\sin(2\pi(mf_0)t)\right)
\end{aligned} \tag{1.3.7}$$

with period P, and the fundamental frequency $f_0 = 1/P$, which has mean and power

$$\mathbf{AV}_x = a_0 \quad \text{and} \quad \mathbf{PW}_x = a_0^2 + \frac{1}{2}\sum_{m=1}^{M}(a_m^2 + b_m^2). \qquad (1.3.8)$$

The above result follows from the fact that not only are sine and cosine signals (of arbitrary frequencies) orthogonal to each other [see (1.3.6)], but also cosines of different frequencies are *orthogonal* to each other, and so are sines. Indeed, if $m \neq n$, that is, $m - n \neq 0$, then

$$\frac{1}{P}\int_0^P \cos(2\pi m f_0 t)\cos(2\pi n f_0 t)\, dt$$

$$= \frac{1}{P}\int_0^P \frac{1}{2}\Big(\cos(2\pi(m-n)f_0 t) + \cos(2\pi(m+n)f_0 t)\Big)\, dt = 0 \quad (1.3.9)$$

and

$$\frac{1}{P}\int_0^P \sin(2\pi m f_0 t)\sin(2\pi n f_0 t)\, dt$$

$$= \frac{1}{P}\int_0^P \frac{1}{2}\Big(\cos(2\pi(m-n)f_0 t) - \cos(2\pi(m+n)f_0 t)\Big)\, dt = 0. \quad (1.3.10)$$

Example 1.3.1 (Superposition of simple cosine oscillations). Consider the signal

$$x(t) = \sum_{m=1}^{12}\frac{1}{m^2}\cos(2\pi m t). \qquad (1.3.11)$$

Its fundamental frequency is $f_0 = 1$, its average is $\mathbf{AV}_x = 0$, and its power is [see (1.3.8)]

$$\mathbf{PW}_x = \frac{1}{2}\sum_{m=1}^{12}\left(\frac{1}{m^2}\right)^2 \approx 0.541.$$

With its sharp cusps, the shape of the above signal is unlike that of any simple harmonic oscillation, and one could start wondering what kind of other periodic signals can be well represented (approximated) by superpositions of harmonic oscillations of the form (1.3.7). The answer, discussed at length in Chap. 2, is that almost all of them can, as long as their power is finite.

The frequency-domain description. The signal $x(t)$ in Example 1.3.1 would be completely specified if, instead of writing the whole formula (1.3.11), we just listed the frequencies present in the signal and the corresponding amplitudes, that is, considered the list

$$(1, 1/1^2), (2, 1/2^2), (3, 1/3^3), \ldots, (12, 1/12^2).$$

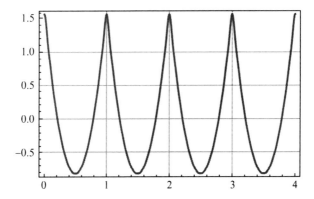

Fig. 1.3.1 The signal $x(t) = \sum_{m=1}^{12} m^{-2} \cos(2\pi m t)$ in its time-domain representation

Similarly, in the case of the general superposition (1.3.7), it would be sufficient to list the cosine and sine frequencies and associated amplitudes, that is, compile the lists

$$(0, a_0), (1f_0, a_1), (2f_0, a_2), \ldots, (Mf_0, a_M) \tag{1.3.12}$$

and

$$(1f_0, b_1), (2f_0, b_2), \ldots, (Mf_0, b_M). \tag{1.3.13}$$

The lists (sequences) (1.3.12) and (1.3.13) are called the *frequency-domain (spectral) representation* of the signal (1.3.7).

Remark 1.3.1 (Amplitude-phase form of the spectral representation). Alternatively, if the signal $x(t)$ in (1.3.7) is rewritten in the amplitude-phase form

$$x(t) = \sum_{m=0}^{M} c_n \cos(2\pi(mf_0)(t + \theta_m)),$$

then the frequency-domain representation must list the frequencies present in the signal, mf_0, $m = 0, 1, \ldots, M$, and the corresponding amplitudes c_m $m = 0, 1, \ldots, M$, and phases θ_m, $m = 0, 1, \ldots, M$.

For the signal from Example 1.3.1, such a representation is graphically pictured in Fig. 1.3.2. We will see in Chap. 2 that, for any periodic signal, the spectrum is always concentrated on a discrete set of frequencies, namely, the multiplicities of the fundamental frequency.

Finally, formula (1.3.8) shows how the total power of the signal $x(t)$ is distributed over different frequencies. Such a distribution, provided by the list

$$(0, a_0^2), (1f_0, (a_1^2 + b_1^2)/2), (2f_0, (a_2^2 + b_2^2)/2), \ldots, (Mf_0, (a_M^2 + b_M^2)/2), \tag{1.3.14}$$

is called the *power spectrum* of the periodic signal (1.3.7).

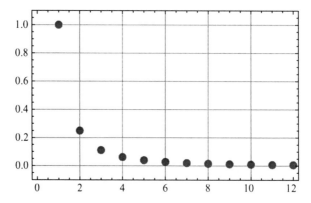

Fig. 1.3.2 The signal $x(t) = \sum_{m=1}^{12} m^{-2} \cos(2\pi m t)$ in its frequency-domain representation. Only the amplitudes of frequencies $m = 1, 2, \ldots, 12$ are shown since all the phase shifts are zero

Observe that, in general, knowledge of the power spectrum is not sufficient for the reconstruction of the signal $x(t)$ itself, while knowledge of the whole representation in the frequency domain is.

To complete our elementary study of periodic signals, note that if an arbitrary signal is studied only in a finite time interval $[0, P]$, then it can always be treated as a periodic signal with period P since one can extend its definition periodically to the whole timeline by copying its waveform from the interval $[0, P]$ to intervals $[P, 2P], [2P, 3P]$, and so on.

1.4 Building a Better Mousetrap: Complex Exponentials

Catching the structure of periodic signals via their decomposition into a superposition of basic trigonometric functions leads to some cumbersome calculations employing various trigonometric identities (as we have seen in Sect. 1.3). A greatly simplified, and also more elegant, approach to the same problem employs a representation of trigonometric functions in terms of exponential functions of the imaginary variable. The cost of moving into the complex domain is not high, as we will rely, essentially, on a single relationship, *j = i   b/c electrical engineers*

$$e^{j\alpha} = \cos\alpha + j \sin\alpha, \quad \text{where} \quad j = \sqrt{-1}, \tag{1.4.1}$$

which is known as *de Moivre's formula*,[4] and which immediately yields two identities:

---

[4] Throughout this book we denote the imaginary unit $\sqrt{-1}$ by the letter j, which is standard usage in the electrical engineering signal processing literature, just as the letter i is reserved for electrical current in the mathematical literature.

$$\cos\alpha = \frac{1}{2}(e^{j\alpha} + e^{-j\alpha}) \quad \text{and} \quad \sin\alpha = \frac{1}{2j}(e^{j\alpha} - e^{-j\alpha}). \quad (1.4.2)$$

In what follows, we are going to routinely utilize the complex number techniques. Thus, for the benefit of the reader, the basic notation and facts about them are summarized in Table 1.4.1.

Table 1.4.1 Complex numbers and de Moivre's formulas

i. By definition,

$$j = \sqrt{-1}.$$

ii. Hence, for any integer m,

$$j^{4m} = 1, \quad j^{4m+1} = j, \quad j^{4m+2} = -1, \quad j^{4m+3} = -j.$$

iii. Cartesian representation of the complex number:

$$z = a + jb, \quad a = \mathrm{Re}\,z, \quad b = \mathrm{Im}\,z,$$

where both a and b are real numbers and are called, respectively, the real and imaginary components of z. The complex number

$$z^* = a - jb$$

is called the *complex conjugate* of z.

iv. The polar representation of the complex number (it is a good idea to think about complex numbers as representing points, or vectors, in the two-dimensional plane spanned by the two basic unit vectors, 1 and j):

$$z = |z|(\cos\theta + j\sin\theta) = |z| \cdot e^{j\theta}$$

and

$$z^* = |z|(\cos\theta - j\sin\theta) = |z| \cdot e^{-j\theta},$$

where

$$|z| = \sqrt{a^2 + b^2} = \sqrt{z \cdot z^*} \quad \text{and} \quad \theta = \mathrm{Arg}\,z = \arctan\frac{\mathrm{Im}z}{\mathrm{Re}z}$$

are called, respectively, the *modulus* of z and the *argument* of z. Alternatively,

$$\mathrm{Re}\,z = \frac{z + z^*}{2} = |z|\cos\theta, \quad \mathrm{Im}\,z = \frac{z - z^*}{2j} = |z|\sin\theta.$$

v. For any complex number $w = \beta + j\alpha$,

$$e^w = e^{\beta + j\alpha} = e^\beta(\cos\alpha + j\sin\alpha).$$

vi. For any complex number $z = a + jb = |z|e^{j\theta}$, and any integer n,

$$z^n = |z|^n e^{jn\theta} = (a^2 + b^2)^{n/2}(\cos n\theta + j\sin n\theta).$$

Since de Moivre's formula is so crucial for us, it is important to understand its origin. The proof is straightforward and relies on the power-series expansion of the exponential function,

$$e^{j\alpha} = \sum_{k=0}^{\infty} \frac{j^k \alpha^k}{k!}.$$ (1.4.3)

However, the powers of the imaginary unit j can be expressed via a simple formula

$$j^k = \begin{cases} 1, & \text{if } k = 4m; \\ j, & \text{if } k = 4m + 1; \\ -1, & \text{if } k = 4m + 2; \\ -j, & \text{if } k = 4m + 3, \end{cases}$$

so the whole series (1.4.3) splits neatly into the real part, corresponding to even indices of the form $k = 2n, n = 0, 1, 2, \ldots$, and the imaginary part, corresponding to the odd indices of the form $k = 2n + 1, n = 0, 1, 2, \ldots$:

$$\sum_{k=0}^{\infty} \frac{j^k \alpha^k}{k!} = \sum_{n=0}^{\infty} \frac{(-1)^n \alpha^{2n}}{(2n)!} + j \sum_{m=0}^{\infty} \frac{(-1)^n \alpha^{2n+1}}{(2n+1)!}.$$

Now, it suffices to recognize in the above formula the familiar power-series expansions for trigonometric functions,

$$\cos \alpha = \sum_{n=0}^{\infty} \frac{(-1)^n \alpha^{2n}}{(2n)!}, \qquad \sin \alpha = \sum_{m=0}^{\infty} \frac{(-1)^n \alpha^{2n+1}}{(2n+1)!},$$

to obtain de Moivre's formula.

Given de Moivre's formulas, which provide a representation of sine and cosine functions via the complex exponentials, we can now rewrite the general superposition of harmonic oscillation:

$$x(t) = a_0 + \sum_{m=1}^{M} a_m \cos(2\pi m f_0 t) + \sum_{m=1}^{M} b_m \sin(2\pi m f_0 t),$$ (1.4.4)

in terms of the complex exponentials

$$x(t) = \sum_{m=-M}^{M} z_m e^{j 2\pi m f_0 t},$$ (1.4.5)

with the real amplitudes a_m and b_m in representations (1.2.4), and the complex amplitudes z_m in the representation (1.2.4), connected by the formulas

$$a_0 = z_0, \qquad a_m = z_m + z_{-m}, \qquad b_m = j(z_m - z_{-m}), \qquad m = 1, 2, \ldots,$$

or, equivalently,

$$z_0 = a_0 \qquad z_m = \frac{a_m - jb_m}{2}, \qquad z_{-m} = \frac{a_m + jb_m}{2}, \qquad m = 1, 2, \ldots,$$

The above relationships show that for the signal of the form (1.4.5) to represent a real-valued signal $x(t)$, it is necessary and sufficient that the paired amplitudes for symmetric frequencies mf_0 and $-mf_0$ be complex conjugates of each other:

$$z_{-m} = z_m^*, \qquad m = 1, 2, \ldots. \tag{1.4.6}$$

However, in the future it will be convenient to consider general complex-valued signals of the form (1.4.5) without the restriction (1.4.6) on its complex amplitudes.

At the first sight, the above introduction of complex numbers and functions of complex-valued variables may seem an unnecessary complication in the analysis of signals. But let us calculate the power of the signal $x(t)$ given by (1.4.5). The need for unpleasant trigonometric formulas disappears as now we need to integrate only exponential functions. Indeed, remembering that $|z|^2 = z \cdot z^*$ now stands for the square of the modulus of a complex number, we have

$$\mathbf{PW}_x = \frac{1}{P} \int_{t=0}^{P} |x(t)|^2 \, dt = \frac{1}{P} \int_{t=0}^{P} \left| \sum_{m=-M}^{M} z_m e^{j2\pi m f_0 t} \right|^2 dt$$

$$= \frac{1}{P} \int_{t=0}^{P} \left(\sum_{m=-M}^{M} z_m e^{j2\pi m f_0 t} \cdot \sum_{k=-M}^{M} z_k^* e^{-j2\pi k f_0 t} \right) dt$$

$$= \frac{1}{P} \sum_{m=-M}^{M} \sum_{k=-M}^{M} z_m z_k^* \int_{t=0}^{P} e^{j2\pi(m-k)f_0 t} \, dt = \sum_{m=-M}^{M} |z_m|^2, \tag{1.4.7}$$

because, for $m - k \neq 0$,

$$\frac{1}{P} \int_{t=0}^{P} e^{j2\pi(m-k)f_0 t} \, dt = \frac{1}{j2\pi(m-k)f_0} e^{j2\pi(m-k)f_0 t} \Big|_{t=0}^{P} = 0, \tag{1.4.8}$$

as the function $e^{j2\pi(m-k)f_0 t} = \cos(2\pi(m-k)f_0 t) + j\sin(2\pi(m-k)f_0 t)$ is periodic with period P, and for $m - k = 0$,

$$\frac{1}{P} \int_{t=0}^{P} e^{j2\pi(m-k)f_0 t} \, dt = 1. \tag{1.4.9}$$

Thus all the off-diagonal terms in the double sum in (1.4.7) disappear. The formulas (1.4.8) and (1.4.9) express mutual orthogonality and normalization of the complex exponential signals,

$$e^{j2\pi m f_0 t}, \qquad m = 0, \pm 1, \pm 2, \ldots, \pm M.$$

In view of (1.4.7), the distribution of the power of the signal (1.4.5) over different multiplicities of the fundamental frequency f_0 can be written as a list with a simple structure:

$$(m f_0, |z_m|^2), \qquad m = 0, \pm 1, \pm 2, \ldots, \pm M. \tag{1.4.10}$$

Remark 1.4.1 (Aperiodic signals). Nonperiodic signals can also be analyzed in terms of their frequency domains, but their spectra are not discrete. We will study them later.

1.5 Problems and Exercises

1.5.1. Find the real and imaginary parts of $(j + 3)/(j - 3)$; $(1 + j\sqrt{2})^3$; $1/(2 - j)$; $(2 - 3j)/(3j + 2)$.

1.5.2. Find the moduli $|z|$ and arguments θ of complex numbers $z = 5$; $z = -2j$; $z = -1 + j$; $z = 3 + 4j$.

1.5.3. Find the real and imaginary components of complex numbers $z = 5 e^{j\pi/4}$; $z = -2 e^{j(8\pi+1.27)}$; $z = -1 e^j$; $z = 3 e^{je}$.

1.5.4. Show that

$$\frac{5}{(1 - j)(2 - j)(3 - j)} = \frac{j}{2} \quad \text{and} \quad (1 - j)^4 = -4.$$

1.5.5. Sketch sets of points in the complex plane (x, y), $z = x + jy$, such that $|z - 1 + j| = 1$; $|z + j| \le 3$; $\text{Re}\,(z^* - j) = 2$; $|2z - j| = 4$; $z^2 + (z^*)^2 = 2$.

1.5.6. Using de Moivre's formulas, find $(-2j)^{1/2}$ and $\text{Re}\,(1 - j\sqrt{3})^{77}$. Are these complex numbers uniquely defined?

1.5.7. Write the signal $x(t) = \sin t + \cos(3t)/3$ from Fig. 1.1.1 as a sum of phase-shifted cosines.

1.5.8. Using de Moivre's formulas, write the signal $x(t) = \sin t + \cos(3t)/3$ from Fig. 1.1.1 as a sum of complex exponentials.

1.5.9. Find the time average and power of the signal $x(t) = -2e^{-j2\pi 4t} + 3e^{-j2\pi t} + 1 - 2e^{j2\pi 3t}$. What is the fundamental frequency of this signal? Plot the distribution of power of $x(t)$ over different frequencies. Write this (complex) signal in terms of cosines and sines. Find and plot its real and imaginary parts.

1.5.10. Using de Moivre's formula, derive the complex exponential representation (1.4.5) of the signal $x(t)$ given by the cosine series representation $x(t) = \sum_{m=1}^{M} c_m \cos(2\pi m f_0 t + \theta_m)$.

1.5.11. Find the time average and power of the signal $x(t)$ from Fig. 1.3.1. Use a symbolic manipulation language such as *Mathematica* or *Matlab* if you like.

1.5.12. Using a computing platform such as *Mathematica*, *Maple*, or *Matlab*, produce plots of the signals

$$x_M(t) = \frac{\pi}{4} + \sum_{m=1}^{M} \left[\frac{(-1)^m - 1}{\pi m^2} \cos mt - \frac{(-1)^m}{m} \sin mt \right],$$

for $M = 0, 1, 2, 3, \ldots, 9$, and $-2\pi < t < 2\pi$. Then produce their plots in the frequency-domain representation. Calculate their power (again, using *Mathematica*, *Maple*, or *Matlab*, if you wish). Produce plots showing how power is distributed over different frequencies for each of them. Write down your observations. What is likely to happen with the plots of these signals as we take more and more terms of the above series, that is, as $M \to \infty$? Is there a limit signal $x_\infty(t) = \lim_{M \to \infty} x_M(t)$? What could it be?

1.5.13. Use the analog-to-digital conversion formula (1.1.1) to digitize signals from Problem 1.5.12 for a variety of sampling periods and resolutions. Plot the results.

1.5.14. Use your computing platform to produce a discrete-time signal consisting of a string of random numbers uniformly distributed on the interval $[0,1]$. For example, in *Mathematica*, the command

```
Table[Random[], {20}]
```

will produce the following string of 20 random numbers between 0 and 1:

```
{0.175245, 0.552172, 0.471142, 0.910891, 0.219577,
0.198173, 0.667358, 0.226071, 0.151935, 0.42048,
0.264864, 0.330096, 0.346093, 0.673217, 0.409135,
0.265374, 0.732021, 0.887106, 0.697428, 0.7723}
```

Use the "random numbers" string as additive noise to produce random versions of the digitized signals from Problem 1.5.12. Follow the example described in Fig. 1.1.3. Experiment with different string lengths and various noise amplitudes. Then center the noise around zero and repeat your experiments.

Chapter 2
Spectral Representation of Deterministic Signals: Fourier Series and Transforms

In this chapter we will take a closer look at the spectral, or frequency-domain, representation of deterministic (nonrandom) signals which was already mentioned in Chap. 1. The tools introduced below, usually called *Fourier*, or *harmonic, analysis* will play a fundamental role later in our study of random signals. Almost all of the calculations will be conducted in the complex form. Compared with working in the real domain, the manipulation of formulas written in the complex form turns out to be simpler and all the tedium of remembering various trigonometric formulas is avoided. All of the results written in the complex form can be translated quickly into results for real trigonometric series expressed in terms of sines and cosines via the familiar de Moivre's formula from Chap. 1, $e^{jt} = \cos t + j \sin t$.

2.1 Complex Fourier Series for Periodic Signals

Any finite-power, complex-valued signal $x(t)$, periodic with period P (say, seconds), can[1] be written in the form of an *infinite* complex Fourier series, meant as a limit (in a sense to be made more precise later), for $M \to \infty$, of finite superposition of complex harmonic exponentials discussed in Sect. 1.4 [see (1.4.5)]:

$$x(t) = \sum_{m=-\infty}^{\infty} z_m e^{j2\pi m f_0 t} = \sum_{m=-\infty}^{\infty} z_m e^{jm\omega_0 t}, \qquad (2.1.1)$$

where $f_0 = \frac{1}{P}$ is the *fundamental frequency* of the signal (measured in Hz = 1/s), and $\omega_0 = 2\pi f_0$ is called the fundamental *angular velocity* (measured in rad/s). The complex number z_m, where m can take values $\ldots, -2, -1, 0, 1, 2, \ldots$, is called the mth Fourier coefficient of signal $x(t)$. Think about it as the amplitude of the harmonic component, with the frequency $m f_0$, of the signal $x(t)$.

---

[1] For mathematical issues related to the feasibility of such a representation, see the discussion in the subsection of this section devoted to the analogy between the orthonormal basis in a 3D space and complex exponentials.

W.A. Woyczyński, *A First Course in Statistics for Signal Analysis*,
DOI 10.1007/978-0-8176-8101-2_2, © Springer Science+Business Media, LLC 2011

In this text we will carry out our calculations exclusively in terms of the fundamental frequency f_0, although one can find in the printed and software signal processing literature sources where all the work is done in terms of ω_0. It is an arbitrary choice, but some formulas are simpler if written in the frequency domain; transitioning from one system to the other is easily accomplished by adjusting various constants appearing in the formulas.

The infinite Fourier series representation (2.1.1) is unique in the sense that two different signals[2] will have two different sequences of Fourier coefficients. The uniqueness is a result of the fundamental property of complex exponentials

$$e_m(t) := e^{j2\pi m f_0 t}, \qquad m = \ldots, -2, -1, 0, 1, 2, \ldots, \qquad (2.1.2)$$

called *orthonormality*:

The scalar product (sometimes also called the inner, or dot, product) of two complex exponentials e_n and e_m is 0 if the exponentials are different, and it is 1 if they are the same. Indeed,

$$\langle e_n, e_m \rangle := \frac{1}{P} \int_0^P e_n(t) e_m^*(t)\, dt$$

$$= \frac{1}{P} \int_0^P e^{j2\pi(n-m)f_0 t}\, dt = \begin{cases} 0, & \text{if } n \neq m; \\ 1, & \text{if } n = m. \end{cases} \qquad (2.1.3)$$

Recall that for a complex number $z = a + jb = |z|e^{j\theta}$ with real component a and imaginary component b, the complex conjugate $z^* = a - jb = |z|e^{-j\theta}$. Sometimes it is convenient to describe the orthonormality using the *Kronecker delta* notation:

$$\delta(n) = \begin{cases} 0, & \text{if } n \neq 0; \\ 1, & \text{if } n = 0. \end{cases}$$

Then, simply,

$$\langle e_m, e_n \rangle = \delta(n - m).$$

Using the orthonormality property, we can directly evaluate the coefficients z_m in the Fourier series (2.1.1) of a given signal $x(t)$ by formally calculating the scalar product of $x(t)$ and $e_m(t)$:

$$\langle x, e_m \rangle = \frac{1}{P} \int_0^P \left(\sum_{n=-\infty}^{\infty} z_n e_n(t) \right) \cdot e_m^*(t)\, dt$$

$$= \sum_{n=-\infty}^{\infty} z_n \frac{1}{P} \int_0^P e_n(t) e_m^*(t)\, dt = z_m, \qquad (2.1.4)$$

---

[2] Meaning that their difference has positive power.

so that we get an explicit formula for the Fourier coefficient of signal $x(t)$:

$$z_m = \langle x, e_m \rangle = \frac{1}{P} \int_0^P x(t) e^{-j2\pi m f_0 t} \, dt. \tag{2.1.5}$$

Thus, the basic Fourier expansion (2.1.1) can now be rewritten in the form of a formal identity:

$$x(t) = \sum_{n=-\infty}^{\infty} \langle x, e_n \rangle e_n(t). \tag{2.1.6}$$

It is worthwhile recognizing that the above calculations on infinite series and interchanges of the order of integration and infinite summations were purely formal; that is, the soundness of the limit procedures was not rigorously established. The missing steps can be found in the mathematical literature devoted to Fourier analysis.[3] For our purposes, suffice it to say that if a periodic signal $x(t)$ has finite power,

$$PW_x = \frac{1}{P} \int_0^P |x(t)|^2 \, dt < \infty, \tag{2.1.7}$$

and the concept of convergence of the functional infinite series (2.1.1) is defined in the right way, then all of the above formal manipulations can be rigorously justified. We will return to this issue at the end of this section. In what follows we will usually consider signals with finite power.

Real-valued signals. Signal $x(t)$ is real-valued if and only if the coefficients z_m satisfy the algebraic condition

$$z_{-m} = z_m^*, \tag{2.1.8}$$

in which case cancellation of the imaginary parts in the Fourier series (2.1.1) occurs. Indeed, under assumption (2.1.8),

$$z_m = |z_m| e^{j\theta_m}, \quad \theta_{-m} = -\theta_m, \tag{2.1.9}$$

and since

$$\frac{e^{j\alpha} + e^{-j\alpha}}{2} = \cos\alpha,$$

we get

$$x(t) = c_0 + \sum_{m=1}^{\infty} c_m \cos(2\pi m f_0 t + \theta_m), \tag{2.1.10}$$

where

$$c_0 = z_0 \quad \text{and} \quad c_m = 2|z_m|, \quad m = 1, 2, \ldots. \tag{2.1.11}$$

---

[3] See, e.g., A. Zygmund, *Trigonometric Series*, Cambridge University Press, Cambridge, UK, 1959.

The power $\mathbf{PW}_x$ of a periodic signal $x(t)$ given by its Fourier series (2.1.1) can also be directly calculated from its Fourier coefficient z_m. Indeed, again calculating formally, we obtain that

$$\mathbf{PW}_x = \frac{1}{P} \int_0^P |x(t)|^2 \, dt = \frac{1}{P} \int_0^P x(t) x^*(t) \, dt$$

$$= \frac{1}{P} \int_0^P \left(\sum_{k=-\infty}^{\infty} z_k e_k(t) \right) \cdot \left(\sum_{m=-\infty}^{\infty} z_m e_m(t) \right)^* dt$$

$$= \sum_{k=-\infty}^{\infty} \sum_{m=-\infty}^{\infty} z_k z_m^* \frac{1}{P} \int_0^P e_k(t) e_m^*(t) \, dt = \sum_{m=-\infty}^{\infty} z_m z_m^*,$$

in view of the orthonormality (2.1.3) of the complex exponentials. The multiplication of the two infinite series was carried out term by term. The resulting relationship

$$\mathbf{PW}_x = \frac{1}{P} \int_0^P |x(t)|^2 \, dt = \sum_{m=-\infty}^{\infty} |z_m|^2 \qquad (2.1.12)$$

is known as *Parseval's formula*. A similar calculation for the scalar product $(1/P) \int_0^P x(t) y^*(t)$ of two different periodic signals, $x(t)$ and $y(t)$, gives an *extended Parseval formula* listed in Table 2.1.1.

Remark 2.1.1 (Distribution of power over frequencies in a periodic signal). Parseval's formula describes how the power $\mathbf{PW}_x$ of the signal $x(t)$ is distributed over different frequencies. The sequence (or its plot)

$$(m f_0, |z_m|^2), \quad m = 0, \pm 1, \pm 2, \ldots, \qquad (2.1.13)$$

is called the *power spectrum* of the signal $x(t)$. Simply stated, it says that the harmonic component of $x(t)$, with frequency $m f_0$, has power $|z_m|^2$ (always a non-negative number!).

Analogy between the orthonormal basis of vectors in the 3D space $\mathbf{R}^3$ and the complex exponentials; the completeness theorem. It is useful to think about the complex exponentials $e_m(t) = e^{2\pi j m f_0 t}, m = \ldots, -1, 0, 1, \ldots$, as an infinite-dimensional version of the orthonormal basic vectors in $\mathbf{R}^3$. In this mental picture the periodic signal $x(t)$ is now thought of as an infinite-dimensional "vector" uniquely expandable into an infinite linear combination of the complex exponentials in the same way a 3D vector is uniquely expandable into a finite linear combination of the three unit coordinate vectors. Table 2.1.1 describes this analogy more fully. Note that the Parseval formula can now be seen just as an infinite-dimensional extension of the familiar Pythagorean theorem.

Table 2.1.1 Analogy between orthogonal expansions in 3D and in the space of periodic signals with finite power

Objects	
3D vectors	*Signals with finite power*
$\vec{x} = (x_1, x_2, x_3)$	$x(t) = \sum_{m=-\infty}^{\infty} z_m e_m(t),$
$\vec{y} = (y_1, y_2, y_3)$	$y(t) = \sum_{m=-\infty}^{\infty} w_m e_m(t),$
Bases	
Unit coordinate vectors	*Complex exponentials*
$\vdots$	$\vdots$
$\vec{e}_1 = (1, 0, 0)$	$e_1(t) = e^{j2\pi f_0 t}$
$\vec{e}_2 = (0, 1, 0)$	$e_2(t) = e^{j2\pi(2f_0)t}$
$\vec{e}_3 = (0, 0, 1)$	$e_3(t) = e^{j2\pi(3f_0)t}$
$\vdots$	$\vdots$
Scalar products	
$\langle \vec{x}, \vec{y} \rangle = \sum_{i=1}^{3} x_i y_i$	$\langle x(t), y(t) \rangle = \frac{1}{P} \int_0^P x(t) y^*(t) dt$
Orthonormality	
$\langle \vec{e}_m, \vec{e}_n \rangle = \delta(n - m)$	$\langle e_m(t), e_n(t) \rangle = \delta(n - m)$
Expansions	
Basis	*Fourier*
$\vec{x} = \sum_{m=1}^{3} \langle \vec{x}, \vec{e}_m \rangle \vec{e}_m$	$x(t) = \sum_{i=-\infty}^{\infty} \langle x, e_m \rangle e_m(t)$
Formulas	
Pythagoras'	*Parseval's*
$\|\vec{x}\|^2 = \sum_{m=1}^{3} x_m^2$	$PW_x = \frac{1}{P} \int_0^P \|x(t)\|^2 dt$
	$= \sum_{m=-\infty}^{\infty} \|z_m\|^2$
Scalar product	*Extended Parseval's*
$\langle \vec{x}, \vec{y} \rangle = \sum_{m=1}^{3} x_m y_m$	$\frac{1}{P} \int_0^P x(t) y^*(t) dt = \sum_{m=-\infty}^{\infty} z_m w_m^*$

So far, the delicate issue of the very feasibility of the Fourier expansion (2.1.1) for any periodic signal with finite power has been left out. Note that in the 3D case, the fact that any vector $\vec{x}$ is representable in the form $x_1\vec{e}_1 + x_2\vec{e}_2 + x_3\vec{e}_3$, where $\vec{e}_1, \vec{e}_2, \vec{e}_3$ are the unit coordinate vectors, is due to the fact that $\vec{e}_1, \vec{e}_2, \vec{e}_3$ is a "maximal" system of orthogonal vectors in 3D; it cannot be further expanded. In other words, if a vector $\vec{e}$ is orthogonal to $\vec{e}_1, \vec{e}_2, \vec{e}_3$, then it must be zero. A similar situation arises if one considers the system of *all* basic harmonic complex exponentials,[4] $e_m(t) = e^{2\pi j m f_0 t}, m = \ldots, -1, 0, 1, \ldots$, in the space of finite-power periodic complex signals with period $P = 1/f_0$. If $x(t)$ is such a signal and $\langle x(t), e_m(t) \rangle = 0$ for all $m = \ldots, -1, 0, 1, \ldots$, then, necessarily, $x(t) = 0$. This fact is known as the *completeness theorem* for complex exponentials, and one can find its proof in any mathematical textbook on harmonic or functional analysis. Removing even one of the complex exponentials from the above system creates an incomplete orthonormal system.

---

[4] Note that the sequence also includes the constant $e_0(t) \equiv 1$.

Examples. Recall that a signal is called *even* if it is symmetric under the change of the direction of time, i.e., if $x(t) = x(-t)$; it is called *odd* if it is antisymmetric under the change of the direction of time, i.e., if $x(t) = -x(-t)$. Not unexpectedly, the real Fourier expansion of a real-valued signal $x(t)$, whose periodic extension to the whole real line is even, i.e., $x(t) = x(-t)$ for all $t \in \mathbf{R}$, will contain only cosine functions (which are even) and, similarly, the real Fourier expansion of an odd real-valued signal $x(t) = -x(-t)$ will contain only sine functions (which are odd). This phenomenon will be illustrated in the following examples.

Example 2.1.1 (Pure cosine expansion of an even rectangular waveform). Consider a rectangular waveform with period P, and amplitude $a > 0$, defined by the formula

$$x(t) = \begin{cases} a, & \text{for } 0 \leq t < P/4; \\ 0, & \text{for } P/4 \leq t < 3P/4; \\ a, & \text{for } 3P/4 \leq t < P. \end{cases}$$

The signal is pictured in Fig. 2.1.1, for the particular values $P = 1$ and $a = 1$.

Calculating the coefficients z_m in the expansion of the signal $x(t)$ into a complex Fourier series is straightforward here: For $m = 0$,

$$z_0 = \frac{1}{P} \int_0^P x(t) e^{-j2\pi 0 t/P} dt = \frac{a}{P} \left(\frac{P}{4} - 0 + P - \frac{3P}{4} \right) = \frac{a}{2}.$$

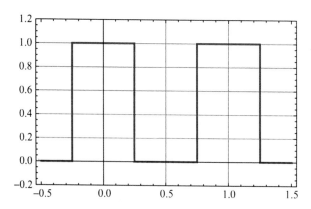

Fig. 2.1.1 An even rectangular waveform signal from Example 2.1.1. The period $P = 1$, and the amplitude $a = 1$

In the case $m \neq 0$,

$$
\begin{aligned}
z_m &= \frac{1}{P} \int_0^P x(t) e^{-j2\pi mt/P} \, dt \\
&= \frac{a}{P} \left(\int_0^{P/4} e^{-j2\pi mt/P} \, dt + \int_{3P/4}^P e^{-j2\pi mt/P} \, dt \right) \\
&= \frac{a}{P} \left(\frac{P}{-j2\pi m} e^{-j2\pi mt/P} \Big|_0^{P/4} + \frac{P}{-j2\pi m} e^{-j2\pi mt/P} \Big|_{3P/4}^P \right) \\
&= \frac{a}{-j2\pi m} \left(e^{-j(\pi/2)m} - 1 - e^{-j(3\pi/2)m} + 1 \right) \\
&= -\frac{a}{\pi m} e^{-j(2\pi/2)m} \left(\frac{e^{j(\pi/2)m} - e^{-j(\pi/2)m}}{2j} \right) \\
&= -\frac{a}{\pi m} \cos \pi m \, \sin \frac{\pi}{2} m = -\frac{a}{\pi m} (-1)^m \sin \frac{\pi}{2} m.
\end{aligned}
$$

If $m = 2k$, then $\sin((\pi/2)m) = 0$ and if $m = 2k + 1$, $k = 0, \pm 1, \pm 2, \ldots$, then $\sin(\pi/2)m = (-1)^k$, which gives, for $k = \pm 1, \pm 2, \ldots$,

$$z_{2k} = 0$$

and

$$z_{2k+1} = \frac{-a}{\pi(2k+1)} (-1)^{2k+1} (-1)^k = \frac{(-1)^k a}{\pi(2k+1)}.$$

Thus, the complex Fourier expansion of the signal $x(t)$ is

$$x(t) = \frac{a}{2} + \frac{a}{\pi} \sum_{k=-\infty}^{\infty} \frac{(-1)^k}{2k+1} e^{j2\pi(2k+1)t/P}.$$

$\cos(2\pi(2k+1)f_0 t)$
$+ j\sin(\cdots)$

Observe that for any $m = \ldots, -1, 0, 1, \ldots$, we have $z_m = z_{-m}$. Pairing up complex exponentials with the exponents of opposite signs, and using de Moivre's formula, we arrive at the real Fourier expansion that contains only cosine functions:

$k=0$ $k=1$

$$x(t) = \frac{a}{2} + \frac{a}{\pi} \left(2\cos(2\pi t/P) - \frac{2}{3} \cos(2\pi 3t/P) + \cdots \right). \quad *\text{imaginary cancels}$$

Example 2.1.2 (Pure sine expansion of an odd rectangular waveform). Consider a periodic rectangular waveform of period P which is defined by the formula

$$x(t) = \begin{cases} a, & \text{for } 0 \leq t < P/4; \\ 0, & \text{for } P/4 \leq t < 3P/4; \\ -a, & \text{for } 3P/4 \leq t < P. \end{cases}$$

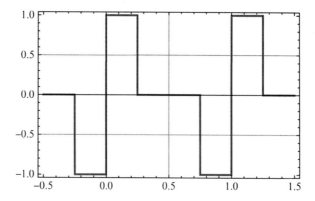

Fig. 2.1.2 An odd rectangular waveform signal from Example 2.1.2. The period $P = 1$, and the amplitude $a = 1$

The signal is pictured in Fig. 2.1.2 for the particular values $P = 1$ and $a = 1$.

For $m = 0$,

$$z_0 = \frac{1}{P} \int_0^P x(t)dt = 0,$$

and for $m \neq 0$,

$$
\begin{aligned}
z_m &= \frac{a}{P}\left(\int_0^{P/4} e^{-j2\pi mt/P} dt - \int_{3P/4}^P e^{-j2\pi mt/P} dt \right) \\
&= \frac{-a}{j2\pi m}\left(e^{-j(\pi/2)m} - 1 - 1 + e^{-j(3\pi/2)m} \right) \\
&= -\frac{a}{j2\pi m}\left[e^{-j(2\pi/2)m}\left(e^{j(\pi/2)m} + e^{-j(\pi/2)m} \right) - 2 \right] \\
&= -\frac{a}{j\pi m}\left((-1)^m \cdot \cos\frac{\pi}{2}m - 1 \right),
\end{aligned}
$$

since, by de Moivre's formula, $e^{-j\pi m} = \cos \pi m - j \sin \pi m$, $\cos \pi m = (-1)^m$, and $\sin \pi m = 0$, for any integer m. On the other hand, $\cos(\pi/2)m = 0$ if m is odd, and $= (-1)^k$ when $m = 2k$ is even, so we get that

$$
z_m = \begin{cases} a/(j\pi(2k+1)), & \text{for odd } m = 2k + 1; \\ a[1 - (-1)^k]/(j\pi 2k), & \text{for even } m = 2k. \end{cases}
$$

Thus, the complex Fourier series of the signal $x(t)$ is of the form

$$
x(t) = \frac{a}{\pi} \sum_{k=-\infty}^{\infty} \left[\frac{1}{j(2k+1)} e^{j2\pi(2k+1)t/P} + \frac{[1 - (-1)^k]}{j2k} e^{j2\pi(2k)t/P} \right].
$$

Observe that in this case, for any $m = \dots, -1, 0, 1, \dots$, we have $z_m = -z_{-m}$, so pairing up the exponentials with opposite signs in the exponents, and using de Moivre's formula, we get a real Fourier series expansion for $x(t)$ that contains only sine functions:

$$x(t) = \frac{2a}{\pi} \left[\sin 2\pi(1)t/P + \sin(2\pi(2)t/P) + \frac{1}{3}\sin(2\pi(3)t/P) \right.$$

$$\left. + 0 \cdot \sin(2\pi(4)t/P) + \frac{1}{5}\sin(2\pi(5)t/P) + \frac{1}{6}\sin(2\pi(6)t/P) + \dots \right].$$

The purpose of going through the above example was to show that, for irregular periodic signals, the calculation of Fourier coefficients can get quite messy although the final result may display a pleasing symmetry.

Example 2.1.3 (A general expansion for a rectangular waveform which is neither odd nor even). Consider a periodic rectangular waveform of period P which is defined by the formula

$$x(t) = \begin{cases} 0, & \text{for } 0 \leq t < P/4; \\ a, & \text{for } P/4 \leq t < P/2; \\ 0, & \text{for } P/2 \leq t < P. \end{cases}$$

The signal is pictured in Fig. 2.1.3 for the parameter values $P = 1$ and $a = 1$ and, for simplicity's sake, we will carry out our calculations only in that case. For $m = 0$,

$$z_0 = \int_{1/4}^{1/2} = \frac{1}{4}.$$

For $m \neq 0$,

$$z_m = |z_m|e^{i\theta_m} = \int_{1/4}^{1/2} e^{-j2\pi mt}\, dt = \frac{1}{-j2\pi m}\left[e^{-j2\pi m/2} - e^{-j2\pi m/4} \right]$$

$$= \frac{1}{\pi m}e^{-j3\pi m/4}\left(\frac{e^{j\pi m/4} - e^{-j\pi m/4}}{2j} \right) == \frac{1}{\pi m}\sin\left(\frac{\pi}{4}m\right)e^{-j3\pi m/4}.$$

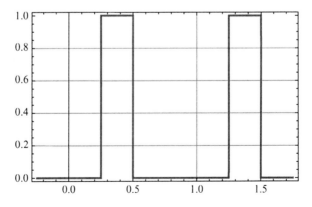

Fig. 2.1.3 A neither odd nor even rectangular waveform signal from Example 2.1.3. The period $P = 1$, and the amplitude $a = 1$

Thus

$$|z_m| = \frac{1}{\pi m} \sin(\pi m/4) \quad \text{and} \quad \theta_m = -j3\pi m/4,$$

and the complex Fourier series for $x(t)$ is

$$x(t) = \frac{1}{4} + \sum_{m=-\infty, m\neq 0}^{\infty} \frac{1}{\pi m} \sin(\pi m/4) e^{-j3\pi m/4} e^{j2\pi mt}.$$

Again, pairing up the complex exponentials with opposite signs in the exponents, we obtain the real expansion in terms of the cosines, but this time with phase shifts that depend on m:

$$x(t) = \frac{1}{4} + \sum_{m=1}^{\infty} \frac{2}{\pi m} \sin(\pi m/4) \cos(2\pi mt - 3\pi m/4),$$

which, using the trigonometric formula $\cos(\alpha + \beta) = \cos\alpha\cos\beta - \sin\alpha\sin\beta$, also can be written as a general real Fourier series,

$$x(t) = a_0 + \sum_{m=1}^{\infty} a_m \cos(2\pi mt) + b_m \sin(2\pi mt),$$

with

$$a_0 = \frac{1}{4}, \qquad a_m = \frac{2}{\pi m} \sin\frac{\pi m}{4} \cos\frac{3\pi m}{4}, \qquad b_m = \frac{2}{\pi m} \sin\frac{\pi m}{4} \sin\frac{3\pi m}{4}.$$

2.2 Approximation of Periodic Signals by Finite Fourier Sums

Up to this point the equality in the Fourier series representation

$$x(t) = \sum_{m=-\infty}^{\infty} \langle x, e_m \rangle e_m(t),$$

for periodic signals, or its real version in terms of sine and/or cosine functions, was understood only formally. But, of course, the usefulness of such an expansion will depend on whether we can show that the signal $x(t)$ can be well approximated by a finite cutoff of the infinite Fourier series, that is, on whether we can prove that

$$x(t) \approx s_M(t) := \sum_{m=-M}^{M} \langle x, e_m \rangle e_m(t) \tag{2.2.1}$$

for M large enough, with the error in the above approximate equality $\approx$ rigorously estimated.

One can pursue several options here:

Approximation in power: Mean-square error. If the error of approximation is measured as the power of the difference between the signal $x(t)$ and the finite Fourier sum $s_M(t)$ in (2.2.1), then the calculation is relatively simple and the error is often called the *mean-square error*. Indeed, using the Parseval formula,

$$
\begin{aligned}
\mathbf{PW}_{x-s_M} &= \frac{1}{P} \int_0^P |x(t) - s_M(t)|^2 \, dt \\[2mm]
&= \frac{1}{P} \int_0^P \left| \sum_{m=-\infty}^{\infty} \langle x, e_m \rangle e_m(t) - s_M(t) \right|^2 dt \\[2mm]
&= \frac{1}{P} \int_0^P \left| \sum_{|m|>M} \langle x, e_m \rangle e_m(t) \right|^2 dt = \sum_{|m|>M} |\langle x, e_m \rangle|^2,
\end{aligned}
$$

which converges to 0 as $M \to \infty$, because we assumed that the power of the signal is finite:

$$
\mathbf{PW}_x = \sum_{m=-\infty}^{\infty} |\langle x, e_m \rangle|^2 < \infty.
$$

Note that the unspoken assumption here is that the orthonormal system $e_n(t)$, $n = 0, \pm 1, \pm 2, \ldots$, is rich enough to make the Fourier representation possible for any finite power signal. This assumption, often called *completeness* of the above orthonormal system, can actually be rigorously proven (see the footnote and other sources cited in the *Bibliographical Comments* at the end of this volume).

Approximation at each time instant t separately. This type of approximation is often called the *pointwise approximation*, and the goal is to verify that, for each time instant t,

$$
\lim_{M \to \infty} s_M(t) = x(t). \tag{2.2.2}
$$

Here the situation is delicate, as examples at the end of this section will show, and the assumption that the signal $x(t)$ has finite power is not sufficient to guarantee the pointwise approximation. Neither is a stronger assumption that the signal is continuous. However,

> *If the signal is continuous, except, possibly, at a finite number of points, and has a bounded continuous derivative, except, possibly, at a finite number of points, then the pointwise approximation (2.2.2) holds true.*

Uniform approximation in time t. If one wants to control the error of approximation simultaneously (uniformly) for all times t, then more stringent assumptions on the signal are necessary. Namely, we have the following theorem:[5]

> *If the signal is continuous everywhere and has a bounded continuous derivative except at a finite number of points, then*

---

[5] Proofs of these two mathematical theorems and other results quoted in this section can be found in, e.g., T. W. Körner, *Fourier Analysis*, Cambridge University Press, Cambridge, UK, 1988.

$$\max_{0 \le t \le P} \left| x(t) - s_M(t) \right| \to 0 \quad \text{as} \quad M \to \infty. \tag{2.2.3}$$

Note that the above statements do not resolve the question of what happens with the finite Fourier sums at discontinuity points of a signal, like those encountered in the rectangular waveforms in Examples 2.1.1–2.1.3. It turns out that under the assumptions of the above-quoted theorems, the points of discontinuity of the signal $x(t)$ are necessarily jumps, that is, the left and right limits

$$x(t_-) = \lim_{s \uparrow t} x(s) \quad \text{and} \quad x(t_+) = \lim_{s \downarrow t} x(s), \tag{2.2.4}$$

exist, and the finite Fourier sums $s_M(x)$ of $x(t)$ converge, as $M \to \infty$, to the average value of the signal at the jump:

$$\lim_{M \to \infty} s_M(t) = \frac{x(t_-) + x(t_+)}{2}. \tag{2.2.5}$$

Example 2.2.1 (Approximation of a rectangular signal by finite Fourier sums). For the signal $x(t)$ in Example 2.1.1, the first three nonzero terms of its cosine expansion were

$$x(t) = \frac{a}{2} + \frac{a}{\pi} \left(2 \cos \left(2\pi \frac{t}{P} \right) - \frac{2}{3} \cos \left(2\pi \frac{3t}{P} \right) + \dots \right).$$

Hence, in the case of period $P = 1$ and amplitude $a = 1$, the first four approximating sums are as follows:

$$s_0(t) = \frac{1}{2}, \qquad s_1(t) = \frac{1}{2} + \frac{2}{\pi} \cos 2\pi t,$$

$$s_2(t) = \frac{1}{2} + \frac{2}{\pi} \cos 2\pi t, \qquad s_3(t) = \frac{1}{2} + \frac{2}{\pi} \cos 2\pi t - \frac{2}{3\pi} \cos 6\pi t.$$

The graphs of $s_1(t)$ and $s_3(t)$ are compared with the original signal $x(t)$ in Figs. 2.2.1 and 2.2.2. Note the behavior of the Fourier sums at the signal's discontinuities, where the Fourier sums converge to the average value of the signal on both sides of the jump according to formula (2.2.5).

Remark 2.2.1 (Irregular behavior of Fourier sums). A word of warning is appropriate here. Abandoning the assumptions in the above two theorems leads very quickly to difficulties with approximating the signal by its Fourier series. For example, there are continuous signals which, at some time instants, have nice Fourier sums diverging to infinity. However, even for them, one can guarantee that the averages of consecutive Fourier sums converge to the signal for each t:

$$\frac{s_0(t) + s_1(t) + \dots + s_M(t)}{M + 1} \to x(t), \quad \text{as} \quad M \to \infty.$$

The expression on the left-hand side of the above formula is called the Mth *Césaro average* of the Fourier series. If one only assumes that the signal $x(t)$ is integrable,

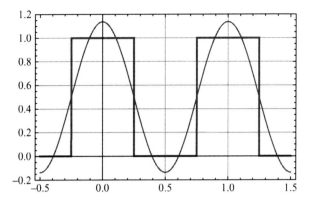

Fig. 2.2.1 Graph of the Fourier sum $s_1(t)$ for the rectangular waveform signal $x(t)$ from Example 2.1.1, plotted against the original signal $x(t)$

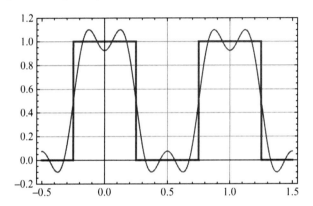

Fig. 2.2.2 Graph of the Fourier sum $s_3(t)$ for the rectangular waveform signal $x(t)$ from Example 2.1.1, plotted against the original signal $x(t)$. Note the behavior of the Fourier sum $s_3(t)$ at the signal's discontinuities, where it matches the average value of the signal at both sides of the jump, reflecting the asymptotics of formula (2.2.5)

that is, $\int_0^P |x(t)|\, dt < \infty$, which is the minimum assumption assuring that the Fourier coefficients $z_m = \langle x, e_m \rangle$ make sense, then one can find signals whose Fourier sums diverge to infinity, for all time instants t.

The Gibbs phenomenon. Another observation is that the finite Fourier sums of a signal satisfying the assumptions of the above-quoted statements, despite being convergent to the signal, may have shapes that are very unlike the signal itself.

Example 2.2.2 (Behavior of Fourier sums at signal's discontinuities). Consider the signal $x(t)$, with period $P = 1$, defined by the formula

$$x(t) = t, \quad \text{for} \quad -1/2 \leq t < 1/2.$$

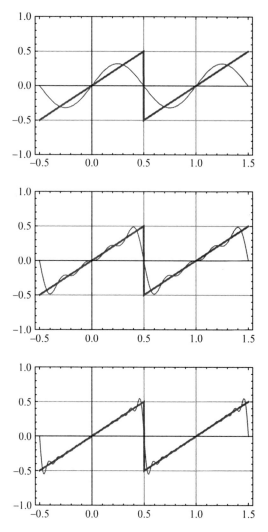

Fig. 2.2.3 Approximation of the periodic signal $x(t)$ from Example 2.2.2 by Fourier sums $s_1(t), s_4(t)$, and $s_{10}(t)$ (*top* to *bottom*). Visible is the Gibbs phenomenon demonstrating that the shape of the Fourier sum near a point of discontinuity of the signal does not necessarily resemble the shape of the signal itself

Clearly, it is an odd signal, so $z_0 = 0$. For $m \neq 0$, integrating by parts,

$$
z_m = \int_{-1/2}^{1/2} te^{-j2\pi mt}\, dt = t\frac{-1}{j2\pi m}e^{-j2\pi mt}\, \Big|_{-1/2}^{1/2} - \frac{-1}{j2\pi m}\int_{-1/2}^{1/2} e^{-j2\pi mt}\, dt
$$

$$
= \frac{-1}{j2\pi m}(-1)^m
$$

because the last integral is zero. The complex Fourier expansion of $x(t)$ is

$$x(t) = \sum_{m=-\infty, m \neq 0}^{\infty} \frac{-1}{j2\pi m} (-1)^m e^{j2\pi mt},$$

which yields a pure sine real Fourier expansion:

$$x(t) = \sum_{m=1}^{\infty} \left(\frac{-1}{j2\pi m}(-1)^m e^{j2\pi mt} + \frac{-1}{j2\pi(-m)}(-1)^{-m} e^{j2\pi(-m)t} \right)$$

$$= \sum_{m=1}^{\infty} \frac{(-1)^{m+1}}{\pi m} \sin(2\pi mt).$$

Figure 2.2.3 shows the approximation of the periodic signal $x(t)$ from Example 2.2.2 by Fourier sums $s_1(t), s_4(t)$, and $s_{10}(t)$. Visible is the *Gibbs phenomenon* demonstrating that the shape of the Fourier sum near a point of discontinuity of the signal does not necessarily resemble the shape of the signal itself. Yet, as the order M of the approximation increases, the oscillations move closer to the jump, so that the mean-square convergence of finite Fourier sums to the signal $x(t)$ is still obtained.

2.3 Aperiodic Signals and Fourier Transforms

Periodic signals with increasing period: From Fourier series to Fourier transform. Consider a signal $x_P(t)$ of period P and fundamental frequency $f_0 = 1/P$. We already know that such signals can be represented by their Fourier series

$$x_P(t) = \sum_{m=-\infty}^{\infty} \left[\frac{1}{P} \int_{-P/2}^{P/2} x(s)e^{-j2\pi mf_0 s} \, ds \right] \cdot e^{j2\pi mf_0 t}. \tag{2.3.1}$$

Notice that, for the purposes of this section, we have written the formula for the Fourier coefficients of $x_P(t)$ as an integral over a symmetric interval $(-P/2, P/2]$ rather than the usual interval of periodicity $(0, P]$. Since the signal $x_P(t)$ and the complex exponentials

$$\exp(-j2\pi mf_0 s) = \cos(2\pi mf_0 s) + j \sin(2\pi mf_0 s)$$

are periodic with period P, any interval of length P will do.

Instead of considering aperiodic signals right off the bat, we will make a gradual transition from the analysis of periodic to aperiodic signals by considering what happens with the Fourier series if, in the above representation (2.3.1), the period P increases to ∞; the limit case of infinite period $P = \infty$ would then correspond to the case of an aperiodic signal.

To see the limit behavior of the Fourier series (2.3.1), we shall introduce the following notation:

1. The multiplicities of the fundamental frequency will become a running discrete variable f_m:

$$f_m = m \cdot f_0;$$

2. The increments of the new running variable will be denoted by

$$\Delta f_m = f_m - f_{m-1} = f_0 = \frac{1}{P}.$$

In this notation the Fourier expansion (2.3.1) can be rewritten in the form

$$x_P(t) = \sum_{m=-\infty}^{\infty} \left[\int_{-P/2}^{P/2} x(s) e^{-j2\pi f_m s} \, ds \right] e^{j2\pi f_m t} \, \Delta f_m \qquad (2.3.2)$$

because $\Delta f_m = f_0 = 1/P$. Now, if the period $P \to \infty$, which is the same as assuming that the fundamental frequency $f_0 = \Delta f_m \to 0$, the sum on the right-hand side of formula (2.3.2) converges to the integral so that our Fourier representation (2.3.2) of a periodic signal $x_P(t)$ becomes the following integral identity for the aperiodic signal:

$$x_\infty(t) = \int_{-\infty}^{\infty} \left[\int_{-\infty}^{\infty} x_\infty(s) e^{-j2\pi f s} \, ds \right] e^{j2\pi f t} \, df. \qquad (2.3.3)$$

The inner transformation,

$$X(f) = \int_{-\infty}^{\infty} x(t) e^{-j2\pi f t} \, dt, \qquad (2.3.4)$$

is called the *Fourier transform* of the signal $x(t)$, and the outer transform,

$$x(t) = \int_{-\infty}^{\infty} X(f) e^{j2\pi f t} \, df, \qquad (2.3.5)$$

is called the *inverse Fourier transform* of the (complex in general) function $X(f)$. The variable in the Fourier transform is the frequency f.

Note that since $|e^{-j2\pi f t}| = 1$, the necessary condition for the existence of the Fourier transform in the usual sense is the absolute integrability of the signal:

$$\int_{-\infty}^{\infty} |x(t)| \, dt < \infty. \qquad (2.3.6)$$

Later on we will try to extend its definition to some important nonintegrable signals.

Example 2.3.1 (Fourier transform of a double exponential signal). Let us trace the above limit procedure in the case of an aperiodic signal $x_\infty(t) = e^{-|t|}$. If this signal is approximated by periodic signals with period P obtained by truncating $x(t)$ to the interval $[-P/2, P/2)$ and extending it periodically, i.e.,

$$x_P(t) = e^{-|t|}, \quad \text{for} \quad -P/2 \leq t < P/2,$$

then the Fourier coefficients of the latter are, remembering that $P = 1/f_0$,

$$z_{m,P} = \frac{1}{P} \int_{-P/2}^{P/2} e^{-|t|} e^{-j2\pi mt/P} \, dt$$

$$= \frac{2f_0}{1 + (2\pi m f_0)^2} \left(1 - e^{-1/(2f_0)} \left(\cos(2\pi m f_0) + 2\pi m f_0 \sin(2\pi m f_0) \right) \right).$$

Since the original periodic signal $x_P(t)$ was even, the Fourier coefficients $z_m = z_{-m}$, so that the discrete spectrum of $x_P(t)$ is symmetric. Now, as $P \to \infty$, that is, $f_0 = 1/P \to 0$, the exponential term $e^{-1/(2f_0)} \to 0$, and with $f_0 = \Delta f, m f_0 = f$, we get that

$$z_{m,P} \to \frac{2}{1 + (2\pi f)^2} \, df, \quad \text{as} \quad P \to \infty.$$

Thus, the Fourier transform of the aperiodic signal $x_\infty(t)$ is

$$X_\infty(f) = \frac{2}{1 + (2\pi f)^2}.$$

Taking the inverse Fourier transform, we verify[6] that

$$\int_{-\infty}^{\infty} \frac{2}{1 + (2\pi f)^2} e^{j2\pi ft} \, df = e^{-|t|}.$$

Figure 2.3.1 illustrates the convergence, as the period P increases, of Fourier coefficients $z_{m,P}$ to the Fourier transform $X_\infty(f)$.

2.4 Basic Properties of the Fourier Transform

The property that makes the Fourier transform of signals so useful is its *linearity*; that is, the Fourier transform of a linear composition $\alpha x(t) + \beta y(t)$ of signals $x(t)$ and $y(t)$ is the same linear composition $\alpha X(f) + \beta Y(f)$ of their Fourier transforms.

---

[6] When faced with integrals of this sort, the reader is advised to consult a book of integrals, or a computer package such as *Mathematica* or *Maple*.

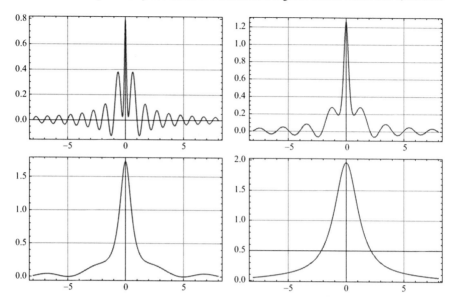

Fig. 2.3.1 Adjusted Fourier coefficients $Z_{m,P} \cdot P$ of a truncated and periodically extended double exponential signal from Example 2.3.1 (shown above, for graphical convenience, as functions of continuous parameter m) approach the Fourier transform $X_\infty(f)$ of the aperiodic signal $x_\infty(t) = e^{-|t|}$. The values of P, from top left to bottom right, are 1, 2, 4, 8

To facilitate notation, we will often denote the fact that $X(f)$ is the Fourier transform of the signal $x(t)$ by writing $x(t) \mapsto X(f)$ [read "$x(t)$ maps into $X(f)$"]. So

$$\alpha x(t) + \beta y(t) \longmapsto \alpha X(f) + \beta Y(f). \tag{2.4.1}$$

The proof is instantaneous using linearity of the integral.

The familiar Parseval formula for periodic signals carries over in the form

$$\mathbf{EN}_x = \int_{-\infty}^{\infty} |x(t)|^2 \, dt = \int_{-\infty}^{\infty} |X(f)|^2 \, df. \tag{2.4.2}$$

It shows how the total energy of the signal is distributed over the *continuous* range of frequencies from minus to plus infinity. The nonnegative function $|X(f)|^2$ is called the *energy spectrum* of the aperiodic signal $x(t)$. The energy of the signal contained in the frequency band $[f_1, f_2]$ can then be calculated as the integral of the square of the modulus of its Fourier transform over that frequency interval:

$$\mathbf{EN}_x[f_1, f_2] = \int_{f_1}^{f_2} |X(f)|^2 \, df. \tag{2.4.2a}$$

An observant reader will see immediately that integrability of the signal necessary to define the Fourier transform is not sufficient for the validity of the Parseval formula

(2.4.2), as the finiteness of the integral $\int_{-\infty}^{\infty} |x(t)| \, dt$ does not imply that the signal has finite energy EN_x (and, vice versa, finiteness of EN_x does not imply the absolute integrability of the signal; see Problem 2.7.11).

Parseval's formula also has the following useful extension:

$$\int_{-\infty}^{\infty} x(t) \cdot y(t) \, dt = \int_{-\infty}^{\infty} X(f) \cdot Y^*(f) \, df. \tag{2.4.3}$$

In the context of transmission of signals through linear systems, the critical property of the Fourier transform is that the *convolution* $[x * y](t)$ of signals $x(t)$ and $y(t)$,

$$[x * y](t) = \int_{-\infty}^{\infty} x(s) y(t - s) \, ds, \tag{2.4.4}$$

a fairly complex, nonlocal operation, has the Fourier transform that is simply the pointwise product of the corresponding Fourier transforms:

$$[x * y](t) \longmapsto X(f) \cdot Y(f). \tag{2.4.5}$$

Indeed,

$$\int_{-\infty}^{\infty} [x * y](t) e^{-j2\pi ft} \, dt = \int_{-\infty}^{\infty} \left[\int_{-\infty}^{\infty} x(s) y(t - s) \, ds \right] e^{-j2\pi ft} \, dt$$

$$= \int_{-\infty}^{\infty} \int_{-\infty}^{\infty} y(t - s) e^{-j2\pi f(t-s)} x(s) e^{-j2\pi fs} \, ds \, dt$$

$$= \int_{-\infty}^{\infty} y(u) e^{-j2\pi fu} \, du \cdot \int_{-\infty}^{\infty} x(s) e^{-j2\pi fs} \, ds$$

$$= X(f) \cdot Y(f),$$

where the penultimate equality resulted from the substitution $t - s = u$.

Since many electrical circuits are described by linear differential equations, the behavior of the Fourier transform under differentiation of the signal is another important issue. Here the calculation is also direct:

$$\int_{-\infty}^{\infty} x'(t) e^{-j2\pi ft} \, dt = x(t) e^{-j2\pi ft} |_{-\infty}^{\infty} + j2\pi f \int_{-\infty}^{\infty} x(t) e^{-j2\pi ft} \, dt$$

$$= 0 + j2\pi f X_Z(f).$$

The first term is 0 because the signal's absolute integrability (remember, we have to assume it to guarantee the existence of the Fourier transform) implies that $x(\infty) = x(-\infty) = 0$. Thus we have a rule:

$$x'(t) \longmapsto (j2\pi f) \cdot X(f). \tag{2.4.6}$$

Table 2.4.1 Properties of the Fourier transform

Signal		Fourier Transform
	Linearity	
$\alpha x(t) + \beta y(t)$	$\longmapsto$	$\alpha X(f) + \beta Y(f)$
	Convolution	
$[x * y](t)$	$\longmapsto$	$X(f) \cdot Y(f)$
	Differentiation	
$x^{(n)}(t)$	$\longmapsto$	$(j 2\pi f)^n X(f)$
	Time reversal	
$x(-t)$	$\longmapsto$	$X(-f)$
	Time delay	
$x(t - t_0)$	$\longmapsto$	$X(f) \cdot e^{-j 2\pi t_0 f}$
	Frequency translation	
$x(t) \cdot e^{j 2\pi f_0 t}$	$\longmapsto$	$X(f - f_0)$
	Frequency differentiation	
$(-j)^n t^n x(t)$	$\longmapsto$	$(2\pi)^{-1} X^{(n)}(f)$
	Frequency convolution	
$x(t) y(t)$	$\longmapsto$	$[X * Y](f)$

Similarly, one can employ the Fourier transform technique to study linear partial differential equations which describe the temporal evolution of physical phenomena in continuous media; see Problem 2.7.18.

The above and other simple-to-derive operational rules for Fourier transforms are summarized in Table 2.4.1.

Example 2.4.1 (Deterministic Gaussian signal and its Fourier transform have the same functional shape). Consider the curious example of a signal of the form $x(t) = e^{-\pi t^2}$, which has the familiar bell shape. Its Fourier transform is

$$X(f) = \int_{-\infty}^{\infty} e^{-\pi t^2 - j 2\pi f t} \, dt = \int_{-\infty}^{\infty} e^{-\pi (t + jf)^2} e^{-\pi f^2} \, dt = e^{-\pi f^2},$$

because

$$\int_{-\infty}^{\infty} e^{-\pi (t + jf)^2} \, dt = \int_{-\infty}^{\infty} e^{-\pi t^2} \, dt = 1.$$

Indeed, changing to polar coordinates r, θ, we can evaluate easily that

$$\left(\int_{-\infty}^{\infty} e^{-\pi t^2} \, dt \right)^2 = \int_{-\infty}^{\infty} e^{-\pi t^2} \, dt \cdot \int_{-\infty}^{\infty} e^{-\pi s^2} \, ds$$

$$= \int_{-\infty}^{\infty} \int_{-\infty}^{\infty} e^{-\pi (t^2 + s^2)} \, dt \, ds = \int_0^{2\pi} d\theta \int_0^{\infty} e^{-\pi r^2} r \, dr = 1.$$

Thus, the signal $x(t) = e^{-\pi t^2}$ has the remarkable property of having the Fourier transform of exactly the same functional shape. This fact has profound consequences in Fourier analysis, mathematical physics, quantum mechanics, and the theory of partial differential equations.

2.5   Fourier Transforms of Some Nonintegrable Signals; Dirac's Delta Impulse

There exist important nonintegrable signals, such as $x(t) = \text{const}$ or $x(t) = \cos t$, that are not absolutely integrable over the whole timeline; as a result, their Fourier transforms are not well defined in the context of the classical calculus. Nevertheless, to cover these and other important cases, it is possible to extend the standard calculus by introduction of the *Dirac delta* "function" $\delta(f)$, which, loosely speaking, is an infinitely high but infinitely narrow spike located at $f = 0$ which, very importantly, has "area," that is, "integral," equal to 1. Of course, one can similarly introduce the time-domain Dirac delta $\delta(t)$, in which case it is often called the *Dirac delta impulse*.

Heuristically (but one can also make this approach rigorous), the best way to think about the Dirac delta is as a limit,

$$\delta(f) = \lim_{\epsilon \to 0} r_\epsilon(f), \qquad (2.5.1)$$

where

$$r_\epsilon(f) = \begin{cases} 1/(2\epsilon), & \text{for } -\epsilon \le f \le +\epsilon; \\ 0, & \text{elsewhere,} \end{cases}$$

is a family of rectangular functions of width 2ϵ, which have area 1 underneath; see Fig. 2.5.1.

Obviously, the choice of the rectangular functions is not unique here. Any sequence of nonnegative functions which integrate to 1 over the whole real line and converge to zero pointwise at every point different from the origin would do. For example, as approximants to the Dirac delta, we can also take the family of double-sided exponential functions of the variable x,

$$\frac{1}{2a} \exp\left(\frac{|f|}{a}\right),$$

indexed by parameter $a \to 0+$. Three functions of this family, for the parameter values $a = 1, 1/3, 1/9$, are pictured in Fig. 2.5.2.

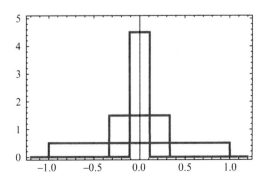

Fig. 2.5.1 Approximation of the Dirac delta $\delta(f)$ by rectangular functions $r_\epsilon(f)$ for $\epsilon = 1, 1/3,$ and $1/9$

Fig. 2.5.2 Approximation of the Dirac delta $\delta(f)$ by two-sided exponential functions $(1/(2a)) \exp(-|f|/a)$ for $a = 1, 1/3$, and $1/9$

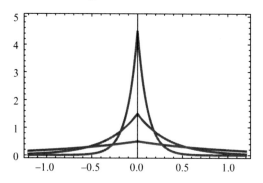

More formally, the Dirac delta will be defined here as a "functional" characterized by its "probing property" describing its scalar products with other, regular functions:

$$\langle \delta, X \rangle := \int_{-\infty}^{\infty} \delta(f) X(f) \, df = X(0). \qquad (2.5.2)$$

In other words, integration of a function $X(f)$ against the Dirac delta produces the value of that function at $f = 0$. This property permits us to use the Dirac delta operationally whenever it appears inside integrals.

The "probing" formula (2.5.2) can be justified by remembering our intuitive definition (2.5.1). Indeed, if the function $X(f)$ is regular enough, then

$$\int_{-\infty}^{\infty} \delta(f) X(f) \, df = \lim_{\epsilon \to 0} \int_{-\infty}^{\infty} r_\epsilon(f) X(f) \, df$$

$$= \lim_{\epsilon \to 0} \frac{1}{2\epsilon} \int_{-\epsilon}^{\epsilon} X(f) \, df = X(0)$$

in view of the fundamental theorem of calculus.

Other properties of the Dirac delta follow immediately. For the Dirac delta shifted to $f = f_0$,

$$\int_{-\infty}^{\infty} \delta(f - f_0) X(f) \, df = X(f_0). \qquad (2.5.3)$$

Also,

$$\int_{-\epsilon}^{\epsilon} \delta(f) \, df = 1 \qquad (2.5.4)$$

and

$$\int_{-\infty}^{\infty} \delta(f) X(f) \, df = 0 \quad \text{if} \quad X(0) = 0. \qquad (2.5.5)$$

The last property is often intuitively stated as follows:

$$\delta(f) = 0, \quad \text{for} \quad f \neq 0. \qquad (2.5.6)$$

Equipped with the Dirac delta technique, we can immediately obtain the Fourier transform of some nonintegrable signals.

Example 2.5.1 (Fourier transforms of complex exponentials). Finding the Fourier transform of the harmonic oscillation signal $x(t) = e^{j2\pi f_0 t}$ is impossible by direct integration, as

$$\int_{-\infty}^{\infty} e^{j2\pi f_0 t} e^{-j2\pi f t} dt$$

$$= \frac{1}{j2\pi(f_0 - f)}\left(\cos 2\pi(f_0 - f)t + j\sin 2\pi(f_0 - f)t\right)\Bigg|_{t=-\infty}^{\infty},$$

and the limits

$$\lim_{t\to\pm\infty} \cos 2\pi(f_0 - f)t \quad \text{and} \quad \lim_{t\to\pm\infty} \sin 2\pi(f_0 - f)t$$

do not exist. But one immediately notices that, in view of (2.5.2), the inverse transform of the shifted Dirac delta is

$$\int_{-\infty}^{\infty} \delta(f - f_0)e^{j2\pi f t} df = e^{j2\pi f_0 t}.$$

Thus, the Fourier transform of $x(t) = e^{j2\pi f_0 t}$ is $\delta(f - f_0)$. In particular, the Fourier transform of a constant 1 is $\delta(f)$ itself.

Example 2.5.2 (Fourier transforms of real harmonic oscillations). The Fourier transform of the signal $x(t) = \cos 2\pi t$ has to be found in a similar fashion, as the direct integration of $\int_{-\infty}^{\infty} \cos(2\pi t) e^{-j2\pi f t} dt$ is again impossible. But one observes that the inverse Fourier transform

$$\int_{-\infty}^{\infty} \frac{1}{2}\Big(\delta(f - 1) + \delta(f + 1)\Big)e^{j2\pi f t} df = \frac{e^{j2\pi t} + e^{-j2\pi t}}{2} = \cos 2\pi t,$$

so the Fourier transform of $\cos 2\pi t$ is $(\delta(f - 1) + \delta(f + 1))/2$.

Table 2.5.1 lists Fourier transforms of some common signals. Here, and thereafter, $u(t)$ denotes Heaviside's unit step function equal to 0, for $t < 0$, and 1, for $t \geq 0$.

Calculus of Dirac delta "functions": Theory of Schwartzian distributions.
There exists a large theory of Dirac delta "functions," and of similar mathematical objects called *distributions* (in the sense of Schwartz),[7] which develops tools that help carry out operations such as distributional differentiation. To give the reader a little taste of it, let us start here with the classical integration-by-parts formula, which, for the usual functions $X(f)$ and $Y(f)$ vanishing at $f = \pm\infty$, states that

---

[7] For a more complete exposition of the theory and applications of the Dirac delta and related "distributions," see A. I. Saichev and W. A. Woyczyński, *Distributions in the Physical and Engineering Sciences, Vol. 1: Distributional Calculus, Integral Transforms and Wavelets*, Birkhäuser Boston, Cambridge, MA, 1997. Also see the Bibliographical Comments at the end of this volume.

Table 2.5.1 Common Fourier transforms

Signal		Fourier Transform
$e^{-a\|t\|}$	$\longmapsto$	$\dfrac{2a}{a^2 + (2\pi f)^2},\quad a > 0$
$e^{-\pi t^2}$	$\longmapsto$	$e^{-\pi f^2}$
$\begin{cases} 1, & \text{for } \|t\| \le 1/2; \\ 0, & \text{for } \|t\| > 1/2. \end{cases}$	$\longmapsto$	$\dfrac{\sin \pi f}{\pi f}$
$\begin{cases} 1 - \|t\|, & \text{for } \|t\| \le 1; \\ 0, & \text{for } \|t\| > 1. \end{cases}$	$\longmapsto$	$\dfrac{\sin^2 \pi f}{\pi^2 f^2}$
$e^{j 2\pi f_0 t}$	$\longmapsto$	$\delta(f - f_0)$
$\delta(t)$	$\longmapsto$	1
$\cos 2\pi f_0 t$	$\longmapsto$	$\dfrac{\delta(f + f_0) + \delta(f - f_0)}{2}$
$\sin 2\pi f_0 t$	$\longmapsto$	$j\,\dfrac{\delta(f + f_0) - \delta(f - f_0)}{2}$
$e^{-at} \cdot u(t)$	$\longmapsto$	$\dfrac{1}{a + j 2\pi f},\quad a > 0$

$$\langle X, Y' \rangle = \int_{-\infty}^{\infty} X(f) \cdot Y'(f)\, df = -\int_{-\infty}^{\infty} X'(f) \cdot Y(f)\, df = -\langle X', Y \rangle. \quad (2.5.7)$$

This identity, applied formally, can be used as the *definition* of the derivative $\delta'(f)$ of the Dirac delta by assigning to it the following probing property:

$$\langle X, \delta' \rangle = \int_{-\infty}^{\infty} X(f) \cdot \delta'(f)\, df = -\int_{-\infty}^{\infty} X'(f) \cdot \delta(f)\, df = -X'(0). \quad (2.5.8)$$

Symbolically, we can write

$$X(f) \cdot \delta'(f) = -X'(f) \cdot \delta(f).$$

In the particular case $X(f) = f$ (here, the function has to be thought of as a limit of functions vanishing at $\pm\infty$), we get

$$f \cdot \delta'(f) = -\delta(f),$$

a useful computational formula.

2.6 Discrete and Fast Fourier Transforms

In practice, for many signals, we only sample the value of the signal at discrete times, although in reality the signal continues between these sampling times. In such cases we can approximate the integrals involved in calculation of the Fourier transforms in the same way as one does in numerical integration in calculus, using left-handed rectangles, trapezoids, Simpson's rule, etc. We use the simplest approximation, which is equivalent to assuming that the signal is constant between the sampling times (and rectangles' areas approximate the area under the function).

So suppose that the sampling period is T_s, with the sampling frequency $f_s = 1/T_s$, so that the signal's sample is given in the form of a finite sequence,

$$x_k = x(kT_s), \quad k = 0, 1, 2, \ldots, N - 1, \tag{2.6.1}$$

so that we can interpret it as a periodic signal with period

$$P = \frac{1}{f_0} = NT_s = \frac{N}{f_s}. \tag{2.6.2}$$

The integral in formula (2.3.1) approximating the Fourier transform of the signal $x(t)$ at discrete frequencies $mf_0, m = 0, 1, 2, \ldots, N - 1$, can now be, in turn, approximated by the sum

$$X_m = X(mf_0) = \frac{1}{P} \sum_{k=0}^{N-1} x(kT_s) e^{-j2\pi mf_0 kT_s} \cdot T_s$$

$$= \frac{1}{N} \sum_{k=0}^{N-1} x_k e^{-j2\pi mk/N}, \tag{2.6.3}$$

in view of relationships (2.6.2). The sequence

$$X_m, \quad m = 0, 1, 2, \ldots, N - 1, \tag{2.6.4}$$

is traditionally called the *discrete Fourier transform (DFT)* of the signal sample $x_k, k = 0, 1, 2, \ldots, N - 1$, described in (2.6.1).

Note that the calculation of the DFT via formula (2.6.3) calls for N^2 multiplications,

$$x_k \cdot e^{-j2\pi mk/N}, \quad m, k = 0, 1, 2, \ldots, N - 1.$$

One often says that the formula's *computational (algorithmic) complexity* is of the order N^2. This computational complexity, however, can be dramatically reduced by cleverly grouping terms in the sum (2.6.3). The technique, which usually is called the *fast Fourier transform (FFT)*, was known to Carl Friedrich Gauss at the beginning of the nineteenth century, but was rediscovered and popularized by Cooley and Tukey in 1965.[8] We will explain it in the special case when the signal's sample size N is a power of 2.

---

[8] J. W. Cooley and O. W. Tukey, "An algorithm for the machine calculation of complex Fourier series," *Math. Comput.* **19**, 297–301, 1965.

So assume that $N = 2^n$, and let $\omega_N = e^{-j2\pi/N}$. The complex number ω_N is called a complex Nth root of unity because $\omega_N^N = 1$. Obviously, for $M = N/2$, we have

$$\omega_{2M}^{(2k)m} = \omega_M^{km}, \quad \omega_M^{M+m} = \omega_M^m, \quad \text{and} \quad \omega_{2M}^{M+m} = -\omega_{2M}^m. \tag{2.6.5}$$

The crucial observation is to recognize that the sum (2.6.3) can be split into two pieces:

$$X_m = \frac{1}{2}\left(X_m^{\text{even}} + X_m^{\text{odd}} \cdot \omega_{2M}^m\right), \tag{2.6.6}$$

where

$$X_m^{\text{even}} = \frac{1}{M}\sum_{k=0}^{M-1} x_{2k}\omega_M^{km}, \quad \text{and} \quad X_m^{\text{odd}} = \frac{1}{M}\sum_{k=0}^{M-1} x_{2k+1}\omega_M^{km}, \tag{2.6.7}$$

and that, in view of (2.6.5),

$$X_{m+M} = \frac{1}{2}\left(X_m^{\text{even}} - X_m^{\text{odd}} \cdot \omega_{2M}^m\right). \tag{2.6.8}$$

As a result, only values X_m, $m = 0, 1, 2, \ldots, M - 1 = N/2 - 1$, have to be calculated by computationally laborious multiplications. The values X_m, $m = M, M + 1, \ldots, 2M - 1 = N - 1$, are simply obtained by formula (2.6.8). The above trick is then repeated at levels $N/2^2, N/2^3, \ldots, 2$. If we denote by $CC(n)$ the *computational complexity* of the above scheme, that is, the number of multiplications required, we see that

$$CC(n) = 2CC(n-1) + 2^{n-1},$$

with the first term on the right being the result of halving the size of the sample at each step, and the second term resulting from multiplications of X_m^{odd} by ω_{2M}^m in (2.6.6) and (2.6.8). Iterating the above recursive relation, one obtains

$$CC(n) = 2^{n-1}\log_2 2^n = \frac{1}{2}N\log_2 N, \tag{2.6.9}$$

a major improvement over the N^2-order of the computational complexity of the straightforward calculation of the DFT.

2.7 Problems and Exercises

2.7.1. Prove that the system of real harmonic oscillations

$$\sin(2\pi m f_0 t), \quad \cos(2\pi m f_0 t), \quad m = 1, 2, \ldots,$$

forms an orthogonal system. Is the system normalized? Is the system complete? Use the above information to derive formulas for coefficients in the Fourier expansions in terms of sines and cosines. Model this derivation on calculations in Sect. 2.1.

2.7.2. Using the results from Problem 2.7.1, find formulas for amplitudes c_m and phases θ_m in the expansion of a periodic signal $x(t)$ in terms of only cosines, $x(t) = \sum_{m=0}^{\infty} c_m \cos(2\pi m f_0 t + \theta_m)$.

2.7.3. Find a general formula for the coefficients c_m in the cosine Fourier expansion for the even rectangular waveform $x(t)$ from Example 2.1.1.

2.7.4. Find a general formula for the coefficients b_m in the sine Fourier expansion for the odd rectangular waveform $x(t)$ from Example 2.1.2.

2.7.5. Carry out calculations of Example 2.1.3 in the case of arbitrary period P and amplitude a.

2.7.6. Find three consecutive approximations by finite Fourier sums of the signal $x(t)$ from Example 2.1.3. Graph them and compare the graphs with the graph of the original signal.

2.7.7. Find the complex and real Fourier series for the periodic signal with period P defined by the formula

$$x(t) = \begin{cases} a, & \text{for } 0 \leq t < P/2; \\ -a, & \text{for } P/2 \leq t < P. \end{cases}$$

In the case $P = \pi$ and $a = 2.5$, produce graphs comparing the signal $x(t)$ and its finite Fourier sums of order 1, 3, and 6.

2.7.8. Find the complex and real Fourier series for the periodic signal with period $P = 1$ defined by the formula

$$x(t) = \begin{cases} 1 - t/2, & \text{for } 0 \leq t < 1/2; \\ 0, & \text{for } 1/2 \leq t < 1. \end{cases}$$

Produce graphs comparing the signal $x(t)$ and its finite Fourier sums of order 1, 3, and 6.

2.7.9. Find the complex and real Fourier series for the periodic signal $x(t) = |\sin t|$. Produce graphs comparing the signal $x(t)$ and its finite Fourier sums of order 1, 3, and 6. in electrical engineering, the signal $|\sin t|$ is produced by running the sine signal through a rectifier.

2.7.10. Find the complex and real Fourier series for the periodic signal with period $P = \pi$ defined by the formula

$$x(t) = e^t, \quad \text{for} \quad -\pi/2 < t \leq \pi/2.$$

Produce graphs comparing the signal $x(t)$ and its finite Fourier sums of order 1, 3, and 6.

2.7.11. Find an example of a signal $x(t)$ that is absolutely integrable, i.e., $\int_{-\infty}^{\infty} |x(t)| \, dt < \infty$, but has infinite energy $EN_x = \int_{-\infty}^{\infty} |x(t)|^2 \, dt$, and vice versa, find an example of a signal which has finite energy but is not absolutely integrable.

2.7.12. Provide a detailed verification of the Fourier transform properties listed in Table 2.4.1. Provide a detailed verification of the Fourier transforms in Table 2.5.1.

2.7.13. (a) The nonperiodic signal $x(t)$ is defined as equal to $1/2$ on the interval $[-1, +1]$ and 0 elsewhere. Plot it and calculate its Fourier transform $X(f)$. Plot the latter.

(b) The nonperiodic signal $y(t)$ is defined as equal to $(t + 2)/4$ on the interval $[-2, 0]$, $(-t + 2)/4$ on the interval $[0, 2]$, and 0 elsewhere. Plot it and calculate its Fourier transform $Y(f)$. Plot the latter.

(c) Compare the Fourier transforms $X(f)$ and $Y(f)$. What conclusion do you draw about the relationship of the original signals $x(t)$ and $y(t)$?

2.7.14. Find the Fourier transform of the periodic signal $x(t) = \sum_{m=-\infty}^{\infty} z_m e^{j2\pi m f_0 t}$.

2.7.15. Find the Fourier transform of the solution $x(t)$ of the differential equation $x''(t) + x(t) = \cos t$.

2.7.16. Find the Fourier transform of the signals given below. Graph both the signal and its Fourier transform (real and imaginary parts separately, if necessary):

(a)
$$x(t) = \frac{1}{1 + t^2}, \quad -\infty < t < \infty.$$

(b)
$$e^{-t^2/2}, \quad -\infty < t < \infty.$$

(c)
$$x(t) = \begin{cases} \sin t \cdot e^{-t}, & \text{for } t \geq 0; \\ 0, & \text{for } t < 0. \end{cases}$$

(d) $x(t) = \sin t \cdot e^{-|t|}$.

(e) $x(t) = y * z(t)$, $y(t) = u(t) - u(t-1)$, $z(t) = e^{-|t|}$

where $u(t)$ is the unit step signal, which $= 0$ for negative t and $= 1$ for $t \geq 0$.

2.7.17. Find the convolution $(x * x)(t)$ if $x(t) = u(t) - u(t-1)$, where $u(t)$ is the unit step function. First, use the original definition of the convolution and then verify your result using the Fourier transform method.

2.7.18. Utilize the Fourier transform (in the space variable z) to find a solution of the diffusion (heat) partial differential equation

$$\frac{\partial u}{\partial t} = \sigma \frac{\partial^2 u}{\partial z^2},$$

for a function $u(t, z)$ satisfying the initial condition $u(0, z) = \delta(z)$. The solution of the above equation is often used to describe the temporal evolution of the density of a diffusing substance.[9]

2.7.19. Assuming the validity of the Parseval formula $\int_{-\infty}^{\infty} |x(t)|^2 \, dt = \int_{-\infty}^{\infty} |X(f)|^2 \, df$, prove its extended version $\int_{-\infty}^{\infty} x(t) \cdot y^*(t) \, dt = \int_{-\infty}^{\infty} X(f) \cdot Y^*(f) \, df$. *Hint:* In the case of real-valued $x(t)$, $y(t)$, $X(f)$, and $Y(f)$, it suffices to utilize the obvious identity $4xy = (x+y)^2 - (x-y)^2$, but in the general, complex case, first verify and then apply the following *polarization identity*:

$$4xy = |x+y|^2 - |x-y|^2 + j(|x+jy|^2 - |x-jy|^2).$$

Remember that the modulus square $|z|^2 = zz^*$.

---

[9] It was the search for solutions to this problem that induced Jean-Baptiste Fourier (born March 21, 1768, in Auxerre, France; died May 16, 1830, in Paris) to introduce in his treatise *Théorie analytique de la chaleur* (1822; *The Analytical Theory of Heat*) the tools of infinite functional series and integral transforms now known under the names of Fourier series and transforms. During the Napoleonic era, Fourier was also known as an Egyptologist and administrator. The modern young author of research papers, impatient with delays in publication of his or her work, should find solace in the fact that the appearance of Fourier's great memoir was held up by the referees for 15 years; it was first presented to the Institut de France on December 21, 1807.

Chapter 3
Random Quantities and Random Vectors

By definition, values of random signals at a given sampling time are random quantities which can be distributed over a certain range of values. The tools for the precise, quantitative description of those distributions are provided by classical *probability theory*. However natural, its development has to be handled with care since the overly heuristic approach can easily lead to apparent paradoxes.[1] But the basic intuitive idea – that for independently repeated experiments, probabilities of their particular outcomes correspond to their relative frequencies of appearance – is correct. Although the concept of probability is more elementary than the concept of cumulative probability distribution function, we assume that the reader is familiar with the former at the high school level, and we start our exposition with the latter, which not only applies universally to all types of data, both discrete and continuous, but also gives us a tool to immediately introduce the probability calculus ideas, including the physically appealing probability density function.

Think here about an electrical engineer whose responsibility is to monitor the voltage on the electrical outlets in the university's circuits laboratory. The record of a month's worth of daily readings on a very sensitive voltmeter may look as follows:

```
109.779, 109.37, 110.733, 109.762, 110.364, 110.73, 109.906,
110.378, 109.132, 111.137, 109.365, 108.968, 111.275, 110.806,
110.99, 111.522, 110.728, 109.689, 111.163, 107.22, 109.661,
108.933, 111.057, 111.055, 112.392, 109.55, 111.042, 110.679,
111.431, 112.06
```

Not surprisingly, the voltage varies slightly and irregularly from day to day, and this variability is visualized in Fig. 3.0.1.

In the presence of such uncertainty, the engineer may want to get a better idea of how the voltage values are distributed within its range and is likely to visualize this information in the form of a histogram, as shown in Fig. 3.0.2.

In this chapter we will discuss analytical tools for the study of such random quantities. The discrete and continuous random quantities are introduced, but we also show that, in the presence of fractal phenomena, the above classification is not exhaustive.

---

[1] See, e.g., Problem 3.7.25.

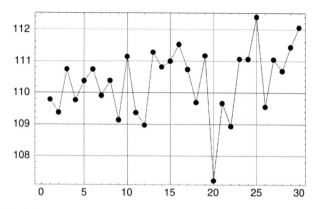

Fig. 3.0.1 Variability of daily voltage readings on an electrical outlet

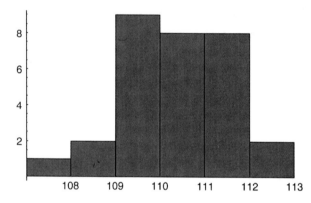

Fig. 3.0.2 The histogram of daily voltage readings on an electrical outlet

3.1 Discrete, Continuous, and Singular Random Quantities

For the purposes of these lectures, *random quantities* (also called *random variables* in the literature), denoted by capital letters X, Y, etc., will symbolize measurements of experiments with uncertain outcomes.

A random quantity X will be fully characterized by its *cumulative distribution function (c.d.f.)*, denoted $F_X(x)$, which gives the probability, $\mathbf{P}(X \leq x)$, the outcomes of experiment X do not exceed the number x:

$$F_X(x) := \mathbf{P}(X \leq x). \tag{3.1.1}$$

Necessarily,

$$F_X(-\infty) = 0, \qquad F_X(\infty) = 1, \tag{3.1.2}$$

the function $F_X(x)$ is nondecreasing,

$$F_X(x) \le F_X(y), \qquad \text{if} \qquad x < y, \tag{3.1.3}$$

and the probability of the measurement being contained in the interval $(a, b]$ is

$$P(a < X \le b) = F_X(b) - F_X(a). \tag{3.1.4}$$

If $a < b < c$, we thus have

$$P(a < X \le c) = F_X(c) - F_X(a) = [F_X(b) - F_X(a)] + [F_X(c) - F_X(b)]$$
$$= P(a < X \le b) + P(b < X \le c).$$

This fundamental property of probabilities, called *additivity*, can be extended from disjoint intervals to more general disjoint[2] sets A and B, yielding the formula

$$P(X \in A \cup B) = P(X \in A) + P(X \in B).$$

In other words, probability behaves like the area measure of planar sets.

Discrete probability distributions. A random quantity X with a discrete probability distribution takes on only (finitely or infinitely many) discrete values, say $x_1, x_2, \ldots$, so that

$$P(X = x_i) = p_i, \quad i = 1, 2, \ldots, \tag{3.1.5a}$$

where

$$0 < p_i < 1, \qquad \sum p_i = 1. \tag{3.1.5b}$$

In the discrete case, the c.d.f. is

$$F_X(x) = \sum_{i=1}^{\infty} p_i u(x - x_i), \tag{3.1.6}$$

where $u(x)$ is the unit step function. In other words, the c.d.f. has jumps of size p_i at locations x_i and is constant at other points of the real line.

Example 3.1.1 (Bernoulli distribution). In this case the values of X, that is, the possible outcomes of the experiment, are assumed to be either 1 or 0 (think about it as a model of an experiment in which "success" or "failure" are the only possible outcomes), with $P(X = 1) = p > 0$, $P(X = 0) = q > 0$, with p, q satisfying condition $p + q = 1$. The c.d.f. of the Bernoulli random quantity is

---

[2] Recall that sets A and B are called *disjoint* if their intersection is the empty set, i.e., $A \cap B = \emptyset$.

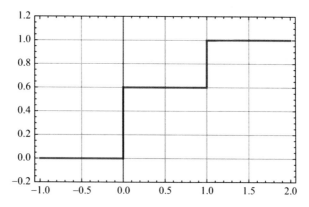

Fig. 3.1.1 The cumulative distribution function $F_X(x)$ of a Bernoulli random quantity X defined in Example 3.1.1, with parameter $p = 0.4$ has a jump of size $q = 1 - 0.4 = 0.6$ at $x = 0$, and a jump of size $p = 0.4$ at $x = 1$

$$F_X(x) = \begin{cases} 0, & \text{for } x < 0; \\ q = 1 - p, & \text{for } 0 \le x < 1; \\ 1, & \text{for } 1 \le x. \end{cases}$$

The Bernoulli family of distributions has one parameter p, which must be a number between 0 and 1. Then $q = 1 - p$. An example is provided in Fig. 3.1.1.

Example 3.1.2 (Binomial distribution). The binomial random quantity X can take values $0, 1, \ldots, n$, with corresponding probabilities

$$p_k = \mathbf{P}(X = k) = \binom{n}{k} p^k (1 - p)^{n-k}, \quad k = 0, 1, 2, \ldots, n,$$

where the binomial coefficient is defined by

$$\binom{n}{k} = \frac{n!}{k!(n-k)!}.$$

Recall, that the name "binomial coefficient" comes from the elementary *binomial formula*

$$(a + b)^n = \sum_{k=0}^{n} \binom{n}{k} a^k b^{n-k},$$

familiar in the special cases:

$$(a + b)^2 = a^2 + 2ab + b^2,$$
$$(a + b)^3 = a^3 + 3a^2b + 3ab^2 + b^3,$$

and so on.

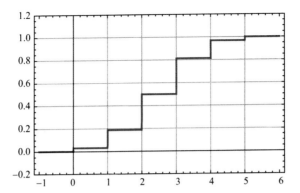

Fig. 3.1.2 The cumulative distribution function $F_X(x)$ of a binomial random quantity X defined in Example 3.1.2, with parameters $p = 0.5$ and $n = 5$

The probabilities $p_k = p_k(n, p)$ in the binomial probability distribution are probabilities that exactly k "successes" occur in n independent[3] Bernoulli experiments, each with probability of "success" equal to p.

The normalization condition $\sum_k p_k = 1$ (3.1.5b) is satisfied here because, in view of the above-mentioned binomial formula,

$$\sum_{k=0}^{n} \binom{n}{k} p^k (1-p)^{n-k} = (p+q)^n = 1.$$

The binomial family of distributions has two parameters: p, which must be between 0 and 1, and n, which can be an arbitrary positive integer. An example is provided in Fig. 3.1.2.

Example 3.1.3 (Poisson distribution). The values of a Poisson random quantity X can be arbitrary nonnegative integers $0, 1, 2, \ldots$, and their probabilities are defined by the formula

$$p_k = \mathbf{P}(X = k) = e^{-\mu} \frac{\mu^k}{k!}, \quad k = 0, 1, 2, \ldots.$$

The normalization condition $\sum_k p_k = 1$ is satisfied in this case because of the power-series expansion for the exponential function:

$$\sum_{k=0}^{\infty} e^{-\mu} \frac{\mu^k}{k!} = e^{-\mu} \sum_{k=0}^{\infty} \frac{\mu^k}{k!} = e^{-\mu} e^{\mu} = 1.$$

The family of Poisson distributions has one parameter $\mu > 0$. Poisson random quantities are often used as models of the number of arrivals of "customers" in

---

[3] A rigorous definition of the concept of the independence of random quantities will be discussed later in this chapter.

queueing systems (an Internet website, a line at the checkout counter, etc.) within a given time interval.

Continuous distributions. A random quantity X is said to have a continuous probability distribution[4] if its c.d.f. $F_X(x)$ can be written as an integral of a certain nonnegative function $f_X(x)$ which traditionally is called the *probability density function* (p.d.f.) of X; that is,

$$F_X(x) = \mathbf{P}(X \le x) = \int_{-\infty}^{x} f_X(z)\,dz. \tag{3.1.7}$$

Then, of course, the probability of the random quantity to assume values between a and b is just the integral of the p.d.f. over the interval $[a, b]$ (see Fig. 3.1.3), where $f_X(x)$ was selected to be $(3/5\sqrt{\pi})e^{-x^2} + (2/5\sqrt{\pi})e^{-(x-2)^2}$. Note that in the continuous case it does not matter whether the interval between a and b is open or closed since the probability of the random quantity taking a particular value is always zero. Thus we have

$$\mathbf{P}(a < X \le b) = F_X(b) - F_X(a) = \int_{a}^{b} f_X(z)\,dz. \tag{3.1.8}$$

Also, necessarily, we have the normalization condition

$$\int_{-\infty}^{\infty} f_X(x)\,dx = F_X(+\infty) = 1, \tag{3.1.9}$$

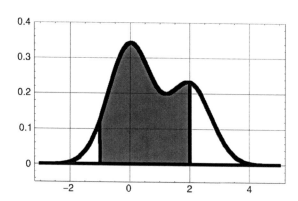

Fig. 3.1.3 The *shaded area* under $f_X(x)$, and above the interval $[-1, 2]$, is equal to the probability that a random quantity X with the p.d.f. $f_X(x)$ takes values in the interval $[-1, 2]$

---

[4] Strictly speaking, c.d.f.s that admit the integral representation (3.1.7), that is, have densities, are called *absolutely continuous distributions*, as there exist continuous c.d.f.s which do not admit this integral representation; see an example of a singular c.d.f. later in this section and, e.g., M. Denker and W. A. Woyczyński, *Introductory Statistics and Random Phenomena: Uncertainty, Complexity and Chaotic Behavior in Engineering and Science*, Birkhäuser Boston, Cambridge, MA, 1998.

and, in view of (3.1.7), and the fundamental theorem of calculus, we can obtain the p.d.f. $f_X(x)$ by differentiation of the c.d.f. $F_X(x)$:

$$\frac{d}{dx} F_X(x) = f_X(x).$$

Example 3.1.4 (Uniform distribution). The density of a uniformly distributed random quantity X is defined to be a positive constant within a certain interval, say $[c, d]$, and zero outside this interval. Thus, because of the normalization condition (3.1.9),

$$f_X(x) = \begin{cases} (d - c)^{-1}, & \text{for } c \le x \le d; \\ 0, & \text{elsewhere.} \end{cases}$$

The family of uniform densities is parameterized by two parameters c and d, with $c < d$. An example is provided in Fig. 3.1.4.

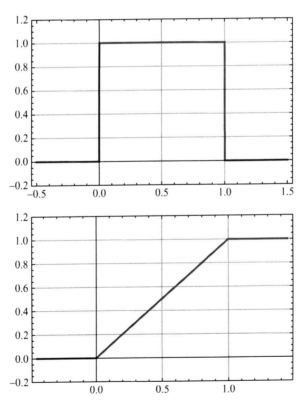

Fig. 3.1.4 (*Top*) The probability density function (p.d.f) $f_X(x)$ for a random quantity with values uniformly distributed over the interval $[0, 1]$. (*Bottom*) The c.d.f. $F_X(x)$ for the same random quantity which was defined in Example 3.1.4

The c.d.f. of a uniform random quantity is

$$F_X(x) = \begin{cases} 0, & \text{for } x < c; \\ (x-c)/(d-c), & \text{for } c \le x \le d; \\ 1, & \text{for } d \le x. \end{cases}$$

Example 3.1.5 (Exponential distribution). An exponentially distributed random quantity X has the p.d.f. of the form

$$f_X(x) = \begin{cases} 0, & \text{for } x < 0; \\ e^{-x/\mu}/\mu, & \text{for } x \ge 0. \end{cases}$$

There is one parameter, $\mu > 0$. The c.d.f. in this case is easily computable:

$$F_X(x) = \begin{cases} 0, & \text{for } x < 0; \\ 1 - e^{-x/\mu}, & \text{for } x \ge 0. \end{cases}$$

An exponential p.d.f. and the corresponding c.d.f. are pictured in Fig. 3.1.5.

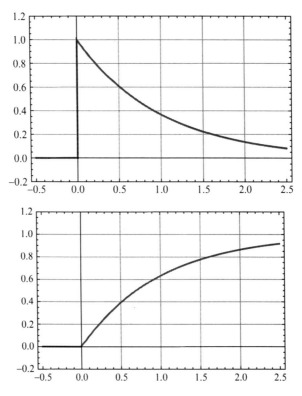

Fig. 3.1.5 (*Top*) The probability density function (p.d.f) $f_X(x)$ of an exponentially distributed random quantity defined in Example 3.1.5, with parameter $\mu = 1$. (*Bottom*) The cumulative distribution function (c.d.f.) $F_X(x)$ for the same random quantity

Exponential p.d.f.s often appear in applications as probability distributions of random waiting times between Poisson events discussed earlier in this section. For example, under certain simplifying assumptions, it can be proven that the random time intervals between consecutive hits at a website have an exponential probability distribution. For this reason, exponential p.d.f.s play a crucial role in analysis of Internet traffic and other queueing networks.

Example 3.1.6 (Gaussian (normal) distribution). The density of a Gaussian (also called normal) random quantity X is defined by the formula

$$f_X(x) = \frac{1}{\sqrt{2\pi}\,\sigma} e^{\frac{-(x-\mu)^2}{2\sigma^2}}.$$

There are two parameters, μ – which is a real number – and $\sigma > 0$, and this distribution is often denoted $N(\mu, \sigma^2)$ p.d.f. (N for "normal"). The Gaussian c.d.f. is of the form (see Fig. 3.1.4)

$$F_X(x) = \int_{-\infty}^{x} \frac{1}{\sqrt{2\pi}\,\sigma} e^{\frac{-(z-\mu)^2}{2\sigma^2}} \, dz,$$

but, unfortunately, the integral cannot be expressed in terms of the elementary functions of the variable x. An example is provided in Fig. 3.1.6. Thus the values of this c.d.f., and the probabilities of a Gaussian random quantity taking values within a given interval, have to be evaluated numerically, using tables (provided at the end of this chapter) or mathematical software such as *Matlab, Maple,* or *Mathematica*; see Example 3.1.6 (continued) ahead.

However, the normalization condition for the Gaussian p.d.f. can be verified directly analytically by a clever trick that replaces the square of the integral by a double integral which is then evaluated in polar coordinates r, θ. We carry out this calculation in the special case $\mu = 0, \sigma^2 = 1$:

$$\left(\int_{-\infty}^{\infty} f_X(x)\, dx \right)^2 = \int_{-\infty}^{\infty} f_X(x)\, dx \cdot \int_{-\infty}^{\infty} f_X(y)\, dy$$

$$= \int_{-\infty}^{\infty}\int_{-\infty}^{\infty} f_X(x) \cdot f_X(y)\, dx\, dy$$

$$= \frac{1}{2\pi} \int_{-\infty}^{\infty}\int_{-\infty}^{\infty} e^{\frac{-x^2-y^2}{2}}\, dx\, dy$$

$$= \frac{1}{2\pi} \int_{0}^{2\pi}\int_{0}^{\infty} e^{\frac{-r^2}{2}} r\, dr\, d\theta = 1.$$

Example 3.1.6 ((continued) Calculations of N (0, 1) probabilities). The values of the Gaussian $N(0, 1)$ cumulative distribution, traditionally denoted $\Phi(x)$, are tabulated

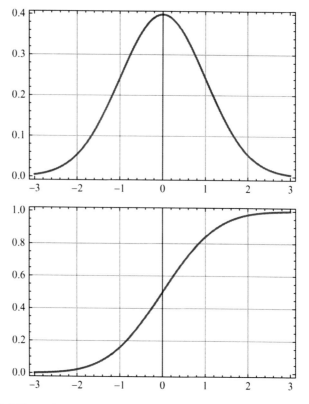

Fig. 3.1.6 (*Top*) The probability density function (p.d.f.) $f_X(x)$ for a Gaussian random quantity defined in Example 3.1.6, with parameters $\mu = 0$, $\sigma = 1$. (*Bottom*) The cumulative distribution function (c.d.f.) $F_X(x)$ for the same random quantity

at the end of this chapter. They are listed only for positive values of the variable x, because, in view of the symmetry of the $N(0, 1)$ density, we have

$$\Phi(-x) = 1 - \Phi(x).$$

Thus

$$\mathbf{P}(-1.53 < X < 2.11) = \Phi(2.11) - \Phi(-1.53) = \Phi(2.11) - (1 - \Phi(1.53))$$
$$= 0.9826 - (1 - 0.9370) = 0.9196.$$

This leaves unanswered the question of how to calculate the general $N(\mu, \sigma^2)$ probabilities. For a solution, see Example 3.1.9.

Remark 3.1.1 (Importance of the Gaussian distribution). The fundamental importance of the Gaussian probability distribution stems from the *central limit theorem* (see Sect. 3.5), which asserts that, for a large number of independent repetitions of experiments with random outcomes, the fluctuations (errors) of the outcomes around their mean value have, approximately, a Gaussian p.d.f. At a more fundamental level, this result is related to the invariance of Gaussian densities under the Fourier transformation; see Example 2.4.1.

Mixed and singular distributions. A random quantity is said to have a c.d.f. of *mixed type* if it has both discrete and continuous components. The c.d.f. thus has both discrete jumps, perhaps infinitely (but countably) many, as well as points of continuous increase where its derivative is well defined. For example, see, Fig. 3.1.7, the c.d.f.

$$F_X(x) = \begin{cases} 0, & \text{for } x < -1; \\ x/6 + 2/6, & \text{for } -1 \le x < 0; \\ x/6 + 4/6, & \text{for } 0 \le x < 1; \\ 1, & \text{for } 1 \le x, \end{cases} \qquad (3.1.10)$$

represents a random quantity X which is uniformly distributed on the intervals $(-1,0) \cup (0,1)$ with probability 1/3, but also takes the discrete values $-1, 0, 1$, with positive probabilities equal to the jump sizes of the c.d.f. at those points. Think here about a cloud of particles randomly and uniformly distributed over the intervals $(-1,0) \cup (0,1)$, with absorbing boundaries at $x = \pm 1$, and a sticky trap at $x = 0$; the probability of finding a particle at those discrete points is positive; 1/6 at $x = \pm 1$, and 1/3 at $x = 0$.

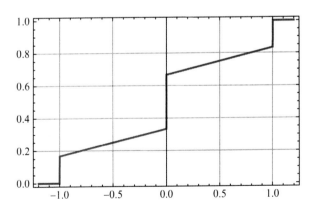

Fig. 3.1.7 The cumulative distribution function (c.d.f.) $F_X(x)$ of mixed type described by formula (3.1.10). This distribution has both discrete and continuous components

Thus, for example,

$$P\left(-\frac{1}{2} < X \le \frac{1}{2}\right) = F_X\left(\frac{1}{2}\right) - F_X\left(-\frac{1}{2}\right) = \left(\frac{1}{12} + \frac{4}{6}\right) - \left(-\frac{1}{12} + \frac{2}{6}\right) = \frac{1}{2},$$

and

$$\mathbf{P}(X = 0) = \lim_{\epsilon \to 0} \mathbf{P}(-\epsilon < X \le \epsilon) = \lim_{\epsilon \to 0} (F_X(\epsilon) - F_X(-\epsilon))$$

$$= \lim_{\epsilon \to 0} [(\epsilon/6 + 4/6) - (-\epsilon/6 + 2/6)] = 1/3.$$

Similarly,

$$\mathbf{P}(X = -1) = 1/6, \quad \mathbf{P}(X = 0) = 2/6, \quad \mathbf{P}(X = 1) = 1/6.$$

Remark 3.1.2 (Mixture of Gaussian p.d.f.s). The reader will notice that the example of a p.d.f. which appeared in Fig. 3.1.3 is a mixture of two Gaussian p.d.f.s.

It is tempting to venture a guess that all c.d.f.s have to be discrete, continuous, or of mixed type. This, however, is not the case.

The limit of the so-called devil's staircase c.d.f.s shown in Fig. 3.1.8 is an example of a c.d.f. which, although continuous and differentiable "almost everywhere," does not have a p.d.f.

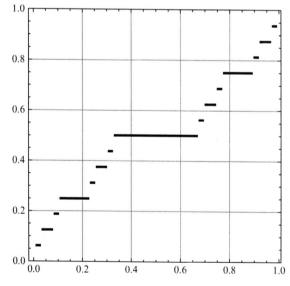

Fig. 3.1.8 The construction of the singular "devil's staircase" c.d.f. $F_X(x)$. It continuously grows from 0, at $x = 0$, to 1, at $x = 1$, and yet it has no density; its derivative is equal to 0 on disjoint intervals whose lengths add up to 1

Observe that inside the interval $[0,1]$ its derivative is 0 on the union of the infinite family of disjoint intervals whose lengths add up to 1. Indeed, as is clear from the construction displayed in Fig. 3.1.8, this set has the linear measure

$$\lim_{n \to \infty} \left(\frac{1}{3} + 2 \cdot \frac{1}{3^2} + \cdots + 2^2 \cdot \frac{1}{3^n} \right) = \frac{1}{3} \sum_{i=0}^{\infty} \left(\frac{2}{3} \right)^i = \frac{1}{3} \cdot \frac{1}{1 - 2/3} = 1,$$

in view of the formula for the sum of a geometric series. Thus integration of this derivative cannot possibly give a c.d.f. which must grow from 0 to 1. Distributions of this type are called *singular*, and they arise in studies of fractal phenomena. One can prove that the set of points of increase of the "devil's staircase" limit, i.e., the set of points on which the probability is concentrated, has a fractional dimension equal to $\ln 2 / \ln 3 = 0.6309\ldots$.[5]

Distributions of functions of random quantities. One often measures random quantities through devices that distort the original quantity X to produce a new random quantity, say, $Y = g(X)$, and the natural question is how the c.d.f. $F_X(x)$ of X is affected by such a transformation. In other words, the question is: Can $F_Y(y)$ be expressed in terms of g and $F_X(x)$? In the case when the transforming function $g(x)$ is *monotonically increasing*, the answer is simple:

$$F_{g(X)}(y) = \mathbf{P}(g(X) \le y) = \mathbf{P}\left(X \le g^{-1}(y) \right) = F_X(g^{-1}(y)), \qquad (3.1.11)$$

where $g^{-1}(y)$ is the inverse function of $g(x)$, that is, $g^{-1}(g(x)) = x$, or, equivalently, if $y = g(x)$, then $x = g^{-1}(y)$.

Remembering the chain rule of the elementary calculus, and the formula for the derivative of the inverse function $g^{-1}(y)$, we also immediately obtain, in the case of monotonically increasing $g(x)$, the *expression of the p.d.f. of $Y = g(X)$ in terms of the p.d.f. of X itself:*

$$f_{g(X)}(y) = \frac{d}{dy} F_X(g^{-1}(y)) = f_X(g^{-1}(y)) \cdot \frac{1}{g'(g^{-1}(y))}. \qquad (3.1.12)$$

Example 3.1.7 (Linear transformation of a standard Gaussian random quantity). Recall that a Gaussian random quantity X is called *standard* [or $N(0, 1)$] if its p.d.f. is of the form

$$f_X(x) = \frac{1}{\sqrt{2\pi}} e^{-\frac{x^2}{2}}.$$

---

[5] See, for example, M. Denker and W. A. Woyczyński, *Introductory Statistics and Random Phenomena: Uncertainty, Complexity and Chaotic Behavior in Engineering and Science*, Birkhäuser Boston, Cambridge, MA, 1998.

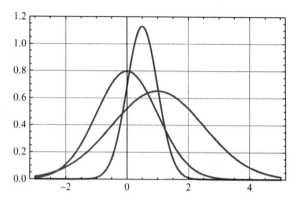

Fig. 3.1.9 Probability density functions of $N(0, 1)$, $N(0.5, 0.25)$, and $N(1, 2.25)$ random quantities (from *left* to *right*)

It is a special case of the general Gaussian p.d.f. introduced in Example 3.1.6, with the parameters μ and σ specified to be 0 and 1, respectively. Consider now a new random quantity Y obtained from X by a linear transformation

$$Y = aX + b, \quad a > 0.$$

Think about this transformation as representing the change in units of measurement and the choice of the origin (like changing the temperature measurements from degrees Celsius to Fahrenheit: If X represents temperature measurements in degrees Celsius, then $Y = 1.8 \cdot X + 32$ gives the same measurements in degrees Fahrenheit).

The transforming function in this case, $y = g(x) = ax + b$, is monotonically increasing, and

$$g'(x) = a \quad \text{and} \quad g^{-1}(y) = (y - b)/a.$$

Formula (3.1.12) now gives the following expression for the p.d.f. of Y:

$$f_Y(y) = \frac{1}{\sqrt{2\pi}} e^{\frac{-((y-b)/a)^2}{2}} \cdot \frac{1}{a} = \frac{1}{\sqrt{2\pi a^2}} e^{\frac{-(y-b)^2}{2a^2}}.$$

The conclusion is that the transformed random quantity Y also has a Gaussian p.d.f., but with parameters $\mu = b$ and $\sigma^2 = a^2$; in other words, Y is $N(b, a^2)$-distributed. Several examples of Gaussian p.d.f.s are shown in Fig. 3.1.9.

Example 3.1.7 ((continued) Calculation of general $N(\mu, \sigma^2)$ probabilities). The relationship established in Example 3.1.7 permits utilization of tables of the $N(0, 1)$ distributions supplied at the end of this chapter to calculate $N(\mu, \sigma^2)$ probabilities for arbitrary values of parameters μ, and $\sigma^2 > 0$. Indeed, if a random quantity Y has the $N(\mu, \sigma^2)$ distribution, then it is of the form

$$Y = \sigma X + \mu,$$

where X has the $N(0, 1)$ distribution with the c.d.f. $F_X(x) = \Phi(x)$, so that

$$F_Y(y) = \mathbf{P}(Y \leq y) = \mathbf{P}(\sigma X + \mu \leq y)$$
$$= \mathbf{P}(X \leq (y - \mu)/\sigma) = \Phi\left(\frac{y - \mu}{\sigma}\right), \qquad (3.1.13)$$

and the values of the latter can be taken from Table 3.6.1. For example, if Y is Gaussian with parameters $\sigma = 1.8$ and $\mu = 32$, then

$$\mathbf{P}(30 < Y < 36) = \Phi\left(\frac{36 - 32}{1.8}\right) - \Phi\left(\frac{30 - 32}{1.8}\right)$$
$$= \Phi(2.22) - (1 - \Phi(-1.11))$$
$$= 0.9868 - (1 - 0.8665) = 0.8533.$$

In the next two examples we will consider the quadratic transformation $Y = X^2/2$ corresponding to calculation of the (random) kinetic energy[6] Y of an object of unit mass $m = 1$, traveling with random velocity X.

Example 3.1.8 (Kinetic energy of a unit mass traveling with random, exponentially distributed velocity). Suppose that the random quantity X has an exponential c.d.f., and the p.d.f. given in Example 3.1.5, with parameter $\mu = 1$. It is transformed by a quadratic "device" $g(x) = x^2/2$ into the random quantity $Y = X^2/2$. Note that the exponential p.d.f. is concentrated on the positive half-line and that the transforming function $g(x)$ is monotonically increasing in that domain. Then the c.d.f. $F_Y(y) = 0$, for $y \leq 0$, and, for $y > 0$, we can repeat the argument from formula (3.1.11) to obtain

$$F_Y(y) = \mathbf{P}(Y \leq y) = \mathbf{P}(X^2/2 \leq y)$$
$$= \mathbf{P}(X \leq \sqrt{2y}) = F_X(\sqrt{2y}) = 1 - e^{-\sqrt{2y}}.$$

Similarly, using (3.1.12), one gets the p.d.f. of $X^2/2$:

$$f_Y(y) = \frac{d}{dy} F_Y(y) = \begin{cases} 0, & \text{for } y \leq 0; \\ e^{-\sqrt{2y}}/\sqrt{2y}), & \text{for } y > 0. \end{cases}$$

Note that this p.d.f. has a singularity at the origin; indeed, $f_Y(y) \uparrow +\infty$ as $y \downarrow 0+$. Observe, however, that the singularity does not affect the p.d.f. normalization condition $\int_{-\infty}^{\infty} f_Y(y)\,dy = 1$.

If the transforming function $y = g(x)$ is *not monotonically increasing* (or decreasing; see Problem 3.7.26 and Sects. 8.1 and 8.2) over the range of the random

---

[6] Recall that an object of mass m traveling with velocity v has kinetic energy $E = mv^2/2$.

quantity X [as, for example, $g(x) = x^2$ in the case when X takes both positive and negative values], then a more subtle analysis is required to find the p.d.f. of the random quantity $Y = g(X)$.

Example 3.1.9 (Square of a standard Gaussian random quantity). Assume that X has the standard $N(0, 1)$ Gaussian p.d.f. and that the transforming function is quadratic: $y = g(x) = x^2$. The quadratic function is monotonically increasing only over the positive half-line; it is monotonically decreasing over the negative half-line. So, we have to proceed with caution, and start with an analysis of the c.d.f. of $Y = X^2$, taking advantage of the symmetry of the Gaussian p.d.f.:

$$F_Y(y) = \mathbf{P}(Y \le y) = \mathbf{P}(X^2 \le y)$$
$$= 2\mathbf{P}(0 \le X \le \sqrt{y}) = 2(F_X(\sqrt{y}) - 1/2).$$

The above formula, obviously, is valid only for $y > 0$; on the negative half-line the c.d.f. of $Y = X^2$ vanishes. Thus the p.d.f. of $Y = X^2$ is

$$f_Y(y) = \frac{d}{dy} F_Y(y) = \begin{cases} 0, & \text{for } y \le 0; \\ e^{-y/2}/(\sqrt{2\pi y}), & \text{for } y > 0. \end{cases}$$

This p.d.f. is traditionally called the *chi-square* probability density function. We'll see its importance in Sect. 3.6, where it plays the central role in the statistical parameter estimation problems.

Random quantities as functions on a sample space. For those who insist on mathematical precision, the above introduction of random quantities via their probability distributions should be preceded by their formal definition as functions on a sample space. This approach had been pioneered by A. N. Kolomogorov[7] and has become a commonly accepted, mainstream approach to mathematical probability theory.

The definition starts with an introduction of the triple $(\Omega, \mathcal{B}, \mathbf{P})$, where the *sample space* Ω is an arbitrary set[8] consisting of *sample points* ω. They should be thought of as labels for different (not necessarily numerical) outcomes of a random experiment being modeled. The *field* $\mathcal{B}$ consists of subsets of the sample space Ω which are called *random events*. To make the logical operations (such as "not," "or," and "and") on random events possible, it is assumed that $\mathcal{B}$ contains the whole

---

[7] See his fundamental *Grundbegriffe der Wahrscheinlichkeisrechnung*, Springer-Verlag, Berlin, 1933, but also an earlier work in the same direction by A. Lomnicki and H. Steinhaus published in the 1923 volume of the journal *Fundamenta mathematicae*, and the Bibliographical Comments at the end of this volume.

[8] Without loss of generality, one can always take as Ω the unit interval $[0, 1]$; see Remark 3.1.2.

sample space Ω and the empty set $\emptyset$, and is closed under complements, unions, and intersections. In other words, one imposes on $\mathcal{B}$ the following *axioms*:

A1.1. $\Omega, \emptyset \in \mathcal{B}$.
A1.2. If $B \in \mathcal{B}$, then its complement $\Omega \setminus B \in \mathcal{B}$.
A1.3. If $A, B \in \mathcal{B}$, then $A \cup B \in \mathcal{B}$, and $A \cap B \in \mathcal{B}$.

The *probability measure* is then defined as a function $\mathbf{P} : \mathcal{B} \mapsto [0, 1]$, assigning to any random event B a real number between 0 and 1, so that it is normalized to 1 on the whole sample space and is additive on mutually exclusive (disjoint) random events. In other words, one imposes on the probability measure the following axioms:

A2.1. $\mathbf{P}(\Omega) = 1$ (normalization).
A2.2. If $A \cap B \in \emptyset$, then $\mathbf{P}(A \cup B) = \mathbf{P}(A) + \mathbf{P}(B)$ (additivity).

Finally, a *random quantity (variable)* X is any function on the sample space Ω which assigns to each sample point ω (that is, to each outcome of a random experiment) a real number $X(\omega)$ in such a way that determining probabilities of $X(\omega)$ taking values in any given interval on the real line is possible. In other words, one demands that the function $X : \Omega \mapsto \mathbf{R}$ is *measurable*, i.e., it satisfies the following axiom:

A3.1. For each $a, b \in \mathbf{R}$, the set of sample points $\{\omega : a < X(\omega) \le b\} \in \mathcal{B}$.

The consequence is that $B = \{\omega : a < X(\omega) \le b\}$ is always a (measurable) random event whose probability $\mathbf{P}(B)$ is well defined. This now permits an introduction of the cumulative distribution function of the random quantity X (and brings us back to the beginning of Sect. 3.1) via the formula

$$F_X(x) = \mathbf{P}(\{\omega : -\infty < X(\omega) \le x\}), \qquad x \in \mathbf{R}.$$

To permit limit operations on random events and random quantities, one usually extends the above axioms to guarantee that infinite unions are permitted in Axioms **A1.3** and **A2.2**. A wide spectrum of examples of sample spaces can encountered in research practice; we provide three – the first is very simple, and the third, rather complex.

Example 3.1.10 (Coin toss – a small sample space). In this case the outcomes can be labeled H (heads) and T (tails), and the sample space, $\Omega = \{H, T\}$, has only two sample points, H and T. The field of random events can be taken to be $\mathcal{B} = \{\emptyset, \{H\}, \{T\}, \Omega\}$. For any number $p \in [0, 1]$, the probability measure $\mathbf{P}$ on all random events in $\mathcal{B}$ can now be defined as follows:

$$\mathbf{P}(\emptyset) = 0, \quad \mathbf{P}(H) = p, \quad \mathbf{P}(T) = 1 - p, \quad \mathbf{P}(\Omega) = 1.$$

Now one can define a variety of random quantities on $(\Omega, \mathcal{B}, \mathbf{P})$. If in the game you are playing one wins \$1 if heads come up and nothing if tails come up, then the corresponding random quantity X is a function on Ω defined by the equalities

$$X(H) = 1, \qquad X(T) = 0,$$

and its probability distribution is

$$\mathbf{P}(\{\omega : X(\omega) = 1\}) = p, \qquad \mathbf{P}(\{\omega : X(\omega) = 0\}) = 1 - p.$$

However, if in the game you are playing one wins \$1 if heads come up and one loses \$1 if tails come up, then the corresponding random quantity X is a function on Ω defined by the equalities

$$X(H) = +1, \qquad X(T) = -1,$$

and its probability distribution is

$$\mathbf{P}(\{\omega : X(\omega) = +1\}) = p, \qquad \mathbf{P}(\{\omega : X(\omega) = -1\}) = 1 - p.$$

Example 3.1.11 (Coin toss – a larger sample space). The above "natural" choice of the "minimal" sample space is not unique. For example, one can choose $\Omega = [0, 1]$, with $\mathbf{P}$ being the length measure of the subsets of the unit interval. Then take

$$X(\omega) = \begin{cases} 0, & \text{for } \omega \in [0, 1 - p]; \\ 1, & \text{for } \omega \in (1 - p, 1). \end{cases}$$

Then, obviously, $\mathbf{P}(\{\omega : X(\omega) = 0\}) = 1 - p$, and $\mathbf{P}(\{\omega : X(\omega) = 1\}) = p$.

Example 3.1.12 (Gas of particles – a large sample space). Consider a gas consisting of $6 \cdot 10^{23}$ (Avogadro's number) of particles (say, of mass 1) moving in $\mathbf{R}^3$ according to Newtonian mechanics. The sample space consists of all possible configurations (states) of the gas described by the particles' positions (x^1, x^2, x^3) and velocities (v^1, v^2, v^3). Hence, each sample point

$$\omega = (x_1^1, x_1^2, x_1^3, v_1^1, v_1^2, v_1^3, \ldots, x_N^1, x_N^2, x_N^3, v_N^1, v_N^2, v_N^3)$$

is a $6 \cdot 6 \cdot 10^{23}$-dimensional vector, and the sample space is of the same huge dimension:

$$\Omega = \mathbf{R}^{6 \cdot 6 \cdot 10^{23}}.$$

The field $\mathcal{B}$ of random events here is also huge and consists of all the subsets of Ω that are defined by imposing upper and lower bounds on the components of the positions and velocities of all $6 \cdot 10^{23}$ particles.

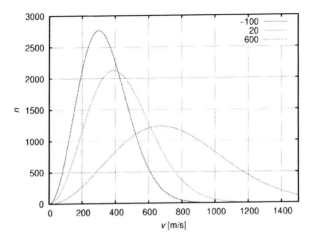

Fig. 3.1.10 Distribution of particle speed for 10^6 oxygen particles at $-100, 20$, and $600°C$ (*left to right*). The speed distribution was derived from the Maxwell–Boltzmann distribution (from http://en.wikipedia.org/wiki/Maxwell-Boltzmann_statistics)

Various probability measures on B can then be defined. In statistical mechanics the standard way to define it is by assigning energy $E(\omega)$ to each configuration ω and then demanding that the probability (fraction of all particles) of the system being in state ω is proportional to $\exp(-\beta)E_\omega$. The resulting probability measure on Ω is called the *Gibbs–Boltzmann measure*.

If the random quantity of interest is just the kinetic energy (temperature) of the configuration,

$$E_\omega = \frac{1}{2} \sum_{i=1}^{N} \left[(v_i^1)^2 + (v_i^2)^2 + (v_i^3)^2 \right],$$

then the above Gibbs–Boltzmann distribution correctly gives the classical Maxwell probability density function of gas particle speed's $s = \sqrt{(v^1)^2 + (v^2)^2 + (v^3)^2}$:

$$f_S(s) = (2/\pi)^{1/2}(kT)^{-3/2}s^2 \exp(-s^2/2kT), \qquad s \geq 0,$$

where k is the Boltzmann constant, and T is the absolute temperature. To accommodate different types of particles, additional parameters are usually included in the above formula; see Fig. 3.1.10 for an example of plots of p.d.f.s of particle speeds for oxygen particles.[9]

Remark 3.1.3 (Unit interval as a universal sample space). For any random quantity X, one can always choose the unit interval $[0, 1]$ as the underlying sample space Ω (although this is not always the most natural selection), with sample points $\omega \in \Omega$ being numbers between 0 and 1. Indeed, equip Ω with the Lebesgue (length)

---

[9] See, e.g., A. H. Carter, *Classical and Statistical Thermodynamics*, Prentice-Hall, Englewood Cliffs, NJ, 2001.

measure as the underlying probability **P**. That is, if $A = [a, b] \subset \Omega = [0, 1]$, then we set

$$\mathbf{P}(A) = \mathbf{P}(\{\omega : a \le \omega \le b\}) = b - a,$$

and let $\omega = F_X(x)$ be the cumulative distribution function of X. Since the above c.d.f. is not necessarily a strictly increasing function, we will define its inverse $F_X^{-1}(\omega)$, $\omega \in [0, 1]$, as the reflection in the diagonal, $x = \omega$, in the (x, ω)-plane of the plot of the p.d.f. $F_X(x)$. More precisely, we uniquely define the (generalized) inverse of the c.d.f. by the equality[10]

$$F_X^{-1}(\omega) = \min\{x : F_X(x) \ge \omega\}.$$

Of course, if $F_X(x)$ is strictly increasing, then the above definition yields the usual inverse function satisfying the conditions

$$F_X^{-1}(F_X(x)) = x \qquad \text{and} \qquad F_X(F_X^{-1}(\omega)) = \omega.$$

In the next step, define

$$X(\omega) = F_X^{-1}(\omega), \qquad \omega \in \Omega = [0, 1].$$

Clearly, $X(\omega)$ defined in this fashion has the correct c.d.f.,

$$\mathbf{P}(\{\omega \in [0, 1] : X(\omega) \le x\}) = \mathbf{P}(\{\omega \in [0, 1] : F_X^{-1}(\omega) \le x\})$$
$$= \mathbf{P}(\{\omega : 0 \le \omega \le F_X(x)\}) = F_X(x).$$

For instance, Example 3.1.1 defines the Bernoulli random quantity as a function on $[0, 1]$ via the above "generalized" inverse of the Bernoulli c.d.f. shown in Fig. 3.1.1. As an example of the strictly increasing c.d.f., we can take the Cauchy random quantity X with the c.d.f.

$$F_X(x) = \frac{1}{\pi}\left(\arctan(x) + \frac{\pi}{2}\right),$$

which continuously increases from 0 to 1 as X ranges from $-\infty$ to $+\infty$. Solving the equation $F_X(x) = \omega$ yields the inverse function

$$F_X^{-1}(\omega) = \tan\left(\pi\omega - \frac{\pi}{2}\right) = X(\omega)$$

and a representation of the Cauchy random quantity as a function on the unit interval. Both the Cauchy c.d.f. and its inverse are shown in Fig. 3.1.11.

---

[10] Traditionally, the inverse of the c.d.f. of a random quantity is called its *quantile function*; see Sect. 3.6.

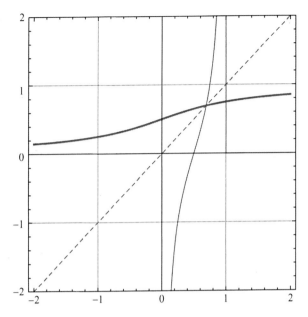

Fig. 3.1.11 The cumulative distribution function, $F_X(x)$, of a Cauchy random quantity (*thick line*) and its inverse, $X(\omega) = F_X^{-1}(\omega)$ (*thin line*), providing a representation of the Cauchy random quantity $X(\omega)$ as a function on the sample space $\Omega = [0, 1]$

3.2 Expectations and Moments of Random Quantities

The *expected value*, or, in brief, the *expectation* of a random quantity X is its mean value (or, for a physics-minded reader, the center of the probability mass) with different values of X given weights equal to their probabilities. The expectation of X will be denoted $\mathbf{E}X$, or $\mathbf{E}(X)$, whichever is more convenient. So, for a discrete random quantity X with $\mathbf{P}(X = x_i) = p_i$, $\sum_i p_i = 1$, we have

$$\mathbf{E}X = \sum_i x_i \, p_i, \tag{3.2.1}$$

and for an (absolutely) continuous random quantity with probability density $f_X(x)$

$$\mathbf{E}X = \int_{-\infty}^{\infty} x f_X(x) \, dx. \tag{3.2.2}$$

More generally, one can consider the expectation of a function $g(X)$ of a random quantity X which is defined by the formulas,

$$\mathbf{E}[\mathbf{g}(\mathbf{X})] = \begin{cases} \sum_i g(x_i) p_i, & \text{in the discrete case;} \\ \int_{-\infty}^{\infty} g(x) f_X(x) \, dx, & \text{in the continuous case.} \end{cases} \tag{3.2.3}$$

In particular, if $g(x) = x^k, k = 1, 2, \ldots$, then the numbers

$$\mu_k(X) = \mathbf{E}g(X) = \mathbf{E}X^k = \begin{cases} \sum_i x_i^k p_i, & \text{in the discrete case;} \\ \int_{-\infty}^{\infty} x^k f_X(x)\, dx, & \text{in the continuous case.} \end{cases}$$

$$(3.2.4)$$

are called k-th moments of X. The first moment $\mu_1 = \mu_1(X)$ is just the expectation of $\mathbf{E}X$ of the random quantity X.

If $g(x) = |x|^\alpha, -\infty < \alpha < \infty$, then

$$m_k(X) = \mathbf{E}|X|^\alpha$$

are called the αth absolute moments, and for $g(x) = |x - \mu_1|^\alpha$, the numbers

$$\mathbf{E}|X - \mu_1|^\alpha = \mathbf{E}|X - \mathbf{E}X|^\alpha$$

are called the αth absolute central moments of X. The latter measure the mean value of the αth power of the deviation of the random quantity X from its expectation $\mathbf{E}X$. In other words, they provide a family of parameters which measure how the values of the random quantity are spread around its "center of mass." In the special case $\alpha = 2$, the second central moment

$$\mathbf{E}(X - \mathbf{E}X)^2 = \begin{cases} \sum_i (x_i - \mu_1)^2 p_i, & \text{in the discrete case;} \\ \int_{-\infty}^{\infty} (x - \mu_1)^2 f_X(x)\, dx, & \text{in the continuous case,} \end{cases}$$

$$(3.2.5)$$

is called the variance of the random quantity X and denoted $\mathrm{Var}(X)$. Again, for a physically minded reader, it is worth noticing that the variance is just the moment of inertia of the probability mass distribution. A simple calculation gives the formula

$$\mathrm{Var}(X) = \mathbf{E}X^2 - (\mathbf{E}X)^2,$$

$$(3.2.6)$$

which is sometimes simpler computationally than (3.2.5); the variance is thus the difference between the second moment (sometimes also called the mean square of a random quantity) and the square of the first moment. This rule is then often phrased: Variance is equal to the mean square minus the squared mean.

Example 3.2.1 (Moments of the Bernoulli distribution). For the Bernoulli random quantity X, with distribution given in Example 3.1.1, all the moments are

$$\mu_k(X) = 1^k \cdot p + 0^k \cdot (1 - p) = p,$$

and the variance is

$$\mathrm{Var}(X) = (1 - p)^2 p + (0 - p)^2 (1 - p) = p(1 - p).$$

Example 3.2.2 (Mean and variance of the uniform distribution). A uniformly distributed random quantity X (see Example 3.1.4) has expectation

$$\mathbf{E}X = \int_c^d x \frac{1}{d-c} dx = \frac{d+c}{2}.$$

Its variance is

$$\mathrm{Var}\,(X) = \int_c^d \left(x - \frac{d+c}{2}\right)^2 \frac{1}{d-c} dx = \frac{(d-c)^2}{12}.$$

Notice that the *expectation*, or *expected value* $\mathbf{E}X$, *of a random quantity* X *scales linearly*, that is,

$$\mathbf{E}(\alpha X) = \alpha \mathbf{E}(X), \qquad -\infty < \alpha < \infty, \tag{3.2.7}$$

so that the change of scale of the measurements affects the expectations proportionally: If, for example, X is measured in meters, then $\mathbf{E}X$ is also measured in meters. Indeed, in the continuous case,

$$\mathbf{E}(\alpha X) = \int_{-\infty}^{\infty} (\alpha x) f_X(x)\, dx = \alpha \int_{-\infty}^{\infty} x f_X(x)\, dx = \alpha \mathbf{E}(X),$$

and the discrete case can be verified in an analogous fashion.

On the other hand, the *variance* $\mathrm{Var}(X)$ *has the quadratic scaling*

$$\mathrm{Var}(\alpha X) = \alpha^2 \mathrm{Var}(X). \tag{3.2.8}$$

This follows immediately from the linear scaling of the expectations (3.2.7) and the formula (3.2.6). Thus the mean-square deviation has a somewhat unpleasant nonlinear property which implies that if X is measured, say, in meters, then its variance is measured in meters squared.

For this reason, one often considers the *standard deviation* $\mathrm{Std}(X)$ of a random quantity X which is defined as the square root of the variance:

$$\mathrm{Std}(X) = \sqrt{\mathrm{Var}\,(X)}. \tag{3.2.9}$$

The standard deviation scales linearly, at least for positive α, since

$$\mathrm{Std}(\alpha X) = |\alpha|\, \mathrm{Std}(X), \qquad -\infty < \alpha < \infty. \tag{3.2.10}$$

This means that changing the measurement units affects the standard deviation proportionately as well. If a random quantity is measured in meters, then its standard deviation is also measured in meters.

Additionally, observe that the *expectation is additive with respect to constants*; that is, for any constant β, $-\infty < \beta < \infty$,

$$\mathbf{E}(X + \beta) = \mathbf{E}(X) + \beta. \tag{3.2.11}$$

The verification is again immediate and follows from the additivity property of the integrals (or, in the discrete case, sums):

$$\mathbf{E}(X + \beta) = \int_{-\infty}^{\infty} (x + \beta) f_X(x)\, dx$$

$$= \int_{-\infty}^{\infty} x f_X(x)\, dx + \int_{-\infty}^{\infty} \beta f_X(x)\, dx = \mathbf{E}(X) + \beta,$$

because $\int_{-\infty}^{\infty} f_X(x)\, dx = 1$.

Finally, the *variance is invariant under translations*; that is, for any constant β, $-\infty < \beta < \infty$,

$$\mathrm{Var}(X + \beta) = \mathrm{Var}(X). \tag{3.2.12}$$

Indeed,

$$\mathrm{Var}(X + \beta) = \mathbf{E}\Big((X + \beta) - \mathbf{E}(X + \beta)\Big)^2 = \mathbf{E}\Big(X + \beta - \mathbf{E}(X) - \beta\Big)^2 = \mathrm{Var}(X).$$

The above properties indicate that any random quantity X can be *standardized* by first centering it, and then rescaling it, so that the standardized random quantity has expectation 0 and variance 1. Indeed, if

$$Z = \frac{X - \mathbf{E}X}{\mathrm{Std}(X)}, \tag{3.2.13}$$

then it immediately follows from (3.2.10) to (3.2.11) that $\mathbf{E}Z = 0$ and $\mathrm{Var}(Z) = 1$.

Example 3.2.3 (Mean and variance of the Gaussian distribution). Let us begin with a random quantity X with the standard $N(0, 1)$ p.d.f. Its expectation is

$$\mathbf{E}(X) = \int_{-\infty}^{\infty} x \frac{1}{\sqrt{2\pi}} e^{-x^2/2}\, dx = 0$$

because the integrand is an odd function and is integrated over the interval $(-\infty, \infty)$, which is symmetric about the origin. Thus its variance is just the second moment (mean square) of X, which can be evaluated easily by integration by parts[11]:

$$\mathrm{Var}(X) = \int_{-\infty}^{\infty} x^2 \frac{1}{\sqrt{2\pi}} e^{-x^2/2}\, dx = \frac{1}{\sqrt{2\pi}} \int_{-\infty}^{\infty} x \cdot (x e^{-x^2/2})\, dx.$$

$$= \frac{1}{\sqrt{2\pi}} \left(-x \cdot e^{-x^2/2} \Big|_{-\infty}^{\infty} + \int_{-\infty}^{\infty} e^{-x^2/2}\, dx \right) = 1,$$

because $\lim_{x \to \pm\infty} x \cdot e^{-x^2/2} = 0$ and $(1/\sqrt{2\pi}) \int_{-\infty}^{\infty} e^{-x^2/2}\, dx = 1$.

---

[11] Recall the integration-by-parts formula: $\int f(x) g'(x)\, dx = f(x) g(x) - \int f'(x) g(x)\, dx$.

Now, let us consider a general Gaussian random quantity Y with $N(\mu, \sigma^2)$ p.d.f.,

$$f_Y(y) = \frac{1}{\sqrt{2\pi\sigma^2}} e^{-\frac{(y-\mu)^2}{2\sigma^2}}.$$

In view of Example 3.1.7,

$$Y = \sigma X + \mu.$$

The above properties of the expectation and the variance [(3.2.7)–(3.2.8) and (3.2.11)–(3.2.12)], immediately give

$$E(Y) = E(\sigma X + \mu) = \sigma E(X) + \mu = \mu$$

and

$$\text{Var}(Y) = \text{Var}(\sigma X + \mu) = \text{Var}(\sigma X) = \sigma^2 \text{Var}(X) = \sigma^2.$$

Thus the parameters μ and σ^2 in the Gaussian $N(\mu, \sigma^2)$ p.d.f. are, simply, its expectation and variance.

Remark 3.2.1 (Sums of random quantities?). Note that the discussions carried out in the previous two sections permitted us, in principle, to determine the probability distributions (and thus expectations, moments, etc.) of functions $g(X)$, once the distribution of X itself was known. However, an effort to determine the distribution of the sum $X + Y$ if the separate distributions of X and Y are known is bound to end up in failure; there is simply not enough information about how the values of X and Y are paired up. This is one of the reasons why one must study the distribution of the pair (X, Y) viewed as the distribution of a single random vector. This will be done in the next section.

3.3 Random Vectors, Conditional Probabilities, Statistical Independence, and Correlations

A *random vector* X has components $X_1, X_2, \ldots, X_d$, which are scalar random quantities; that is,

$$X = (X_1, X_2, \ldots, X_d),$$

where d is the dimension of the random vector. For simplicity of notation, we shall consider first the case of dimension $d = 2$, and we shall write $X = (X, Y)$.

Statistical properties of random vectors are characterized by their *joint probability distributions*. In the discrete case, for a random vector X taking discrete values $x = (x, y)$, the joint probability distribution is

$$P(X = x) = P(X = x, Y = y) = p_X(x, y), \tag{3.3.1}$$

and

$$\sum_{(x,y)} p_X(x, y) = 1. \tag{3.3.2}$$

Example 3.3.1 (A Bernoulli random vector). The random vector (X, Y) takes values $(0,0), (0, 1), (1,0), (1, 1)$, with the following joint probabilities:

$$p_{(X,Y)}(0,0) = (1 - p)^2, \qquad p_{(X,Y)}(0, 1) = p(1 - p),$$
$$p_{(X,Y)}(0, 1) = (1 - p)p, \qquad p_{(X,Y)}(1, 1) = p^2.$$

It is easy to check that

$$\sum_{x=0}^{1} \sum_{y=0}^{1} p_{(X,Y)}(x, y) = 1.$$

In the special case $p = 1/2$, all four possible values of this random vector are taken with the same probability equal to $1/4$.

A continuous random vector is characterized by its *joint p.d.f.* $f_{(X,Y)}(x, y)$, which is a nonnegative function of two variables x, y, such that

$$\int_{-\infty}^{\infty} \int_{-\infty}^{\infty} f_{(X,Y)}(x, y) \, dx \, dy = 1. \tag{3.3.3}$$

In this case the probability that the random vector (X, Y) takes values in a certain domain A of the 2D space is calculated by evaluating the double integral of the joint p.d.f. over the domain A:

$$\mathbf{P}((X, Y) \in A) = \int \int_A f_{(X,Y)}(x, y) \, dx \, dy. \tag{3.3.4}$$

For example, if the domain A is a rectangle $[a, b] \times [c, d] = \{(x, y) : a \le x \le b, c \le y \le d\}$, then

$$\mathbf{P}((X, Y) \in A) = \mathbf{P}(a \le X \le b, c \le Y \le d) = \int_a^b \int_c^d f_{(X,Y)}(x, y) \, dy \, dx. \tag{3.3.5a}$$

If the domain $B = \{(x, y) : x^2 + y^2 \le R^2\}$ is a centered disk of radius R, then

$$\mathbf{P}((X, Y) \in B) = \mathbf{P}(X^2 + Y^2 \le R^2) = \int_{-R}^{R} \int_{-\sqrt{R^2-x^2}}^{\sqrt{R^2-x^2}} f_{(X,Y)}(x, y) \, dy \, dx. \tag{3.3.5b}$$

The graph of a 2D joint p.d.f. is a surface over the (x, y)-plane such that the volume underneath it is equal to 1; see (3.3.3).

Example 3.3.2 (A 2D Gaussian random vector). An example of the 2D Gaussian joint p.d.f. is given by the formula

$$f_{(X,Y)}(x, y) = \frac{1}{2\pi\sigma_x\sigma_y} \exp\left[-\frac{(x - \mu_x)^2}{2\sigma_x^2} - \frac{(y - \mu_y)^2}{2\sigma_y^2}\right], \tag{3.3.6}$$

where $\sigma_x, \sigma_y > 0$, and μ_x, μ_y are arbitrary real numbers. Figure 3.3.1 shows the plot of the surface representing a 2D Gaussian joint p.d.f. in the case $\sigma_x, \sigma_y = 1$ and $\mu_x, \mu_y = 0$.

Calculation of the probabilities $\mathbf{P}(a \leq X \leq b, c \leq Y \leq d)$ is reduced here to the calculation of one-dimensional Gaussian probabilities since the joint 2D density in this case is the product of two 1D Gaussian densities – one depending only on x, and the other on y[12] – and the double integral splits into the product of two single integrals. To obtain numerical values, tables of (or software for) 1D $N(0, 1)$ c.d.f.s have to be used; see Sect. 3.5.

In the special case of equal variances $\sigma_x^2 = \sigma_y^2 = \sigma^2$, the probability that the above Gaussian random vector takes values in a disk of radius R centered at (μ_x, μ_y) can, however, be carried out explicitly by calculation of the integral in polar coordinates (θ, r):

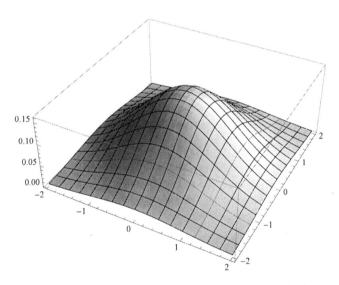

Fig. 3.3.1 Plot of the surface representing a 2D Gaussian joint p.d.f. (3.3.6) in the case $\sigma_x, \sigma_y = 1$ and $\mu_x, \mu_y = 0$

---

[12] We will have more to say about joint p.d.f.s of this type in the next few pages. The multiplicative property is equivalent to the concept of statistical independence of components of a random vector.

$$\mathbf{P}\left((X - \mu_x)^2 + (Y - \mu_y)^2 \le R^2\right)$$

$$= \int_{-R}^{R} \int_{-\sqrt{R^2-x^2}}^{\sqrt{R^2-x^2}} \frac{1}{2\pi\sigma^2} \exp\left[-\frac{x^2+y^2}{2\sigma^2}\right] dy\, dx$$

$$= \frac{1}{2\pi\sigma^2} \int_{0}^{2\pi} \int_{0}^{R} \exp\left[-\frac{r^2}{2\sigma^2}\right] r\, dr\, d\theta$$

$$= \frac{1}{\sigma^2}\left[-\sigma^2 \exp\left[-\frac{r^2}{2\sigma^2}\right]\right]_{0}^{R} = 1 - e^{-R^2/2\sigma^2}.$$

Because the joint p.d.f. gives complete information about the random vector (X, Y), it also yields complete information about the probability distributions of each of the component random quantities. These distributions are called *marginal distributions* of the random vector.

In particular, for a discrete random vector, the marginal distribution of the component X is

$$p_X(x) = \sum_{y} p_{(X,Y)}(x, y). \tag{3.3.7}$$

To find the probability of X taking a particular value x_0, we simply need to sum, over all possible y's, the probabilities of (X, Y) taking values (x_0, y). For a continuous random vector, the marginal p.d.f. of the component X is

$$f_X(x) = \int_{-\infty}^{\infty} f_{(X,Y)}(x, y)\, dy. \tag{3.3.8}$$

It is important to observe that the marginal distributions of components of a random vector *do not* determine its joint distribution. Indeed, the example provided below shows that it is quite possible for random vectors to have the same marginal probability distributions of their components while their joint probability distributions are different.

Example 3.3.3 (Different random vectors with the same marginal probability distributions). A random vector (X, Y) has components X and Y which take values 1, 2, and 3, and 1 and 2, respectively. The joint probability distribution of this random vector is given in the following table:

$Y \backslash X$	1	2	3	Y
1	30/144	24/144	18/144	6/12
2	30/144	24/144	18/144	6/12
X	5/12	4/12	3/12	$\sum = 1$

So, for example, $\mathbf{P}((X, Y) = (3, 2)) = 3/24$. The last row in the above table gives the marginal probability distribution for the component X, and the last column gives the marginal probability distribution for the component Y.

Consider now another random vector (W, Z) with components W and Z which also take values 1, 2, and 3, and 1 and 2, respectively. The joint distribution of this random vector is given by a different table:

$Z \backslash W$	1	2	3	Z
1	1/12	2/12	3/12	6/12
2	4/12	2/12	0	6/12
W	5/12	4/12	3/12	$\sum = 1$

This time $P((X, Y) = (3, 2)) = 0$. The last row in the above table gives the marginal probability distribution for the component W, and the last column gives the marginal probability distribution for the component Z. The marginal probability distributions for vectors (X, Y) and (W, Z) are the same, while their joint distributions are different.

Conditional probabilities. Knowledge of the joint p.d.f. permits us also to introduce the concept of the conditional probability (in the discrete case) and the conditional density (in the continuous case). Thus, the conditional probability of the component X taking value x, given that the second component Y took value y, is given by the formula[13]

$$p_{X|Y}(x|y) \equiv \mathbf{P}(X = x|Y = y) = \frac{P(X = x, Y = y)}{P(Y = y)} = \frac{p_{(X,Y)}(x, y)}{p_Y(y)}, \quad (3.3.9)$$

and the conditional probability density function of X given $Y = y$ is given by the formula

$$f_{X|Y}(x|y) = \frac{f_{(X,Y)}(x, y)}{f_Y(y)}. \quad (3.3.10)$$

In other words, conditional probability distributions are distributions of values of one component of a random vector calculated under the assumption that the value of the other component has already been determined.

Conditional probabilities are *bona fide* probabilities, as they satisfy the normalization property. Indeed, say, in the continuous case, for each fixed y,

$$\int_{-\infty}^{\infty} f_{X|Y}(x|y)\, dx = \frac{\int_{-\infty}^{\infty} f_{(X,Y)}(x, y)\, dx}{f_Y(y)} = \frac{f_Y(y)}{f_Y(y)} = 1,$$

in view of formula (3.3.8), which calculates the marginal density from the joint density.

If the component X of a random vector (X, Y) takes on distinct values $x_1, x_2, \ldots, x_n$, then the additive property of probabilities immediately gives the following *total probability formula:*

---

[13] The notation $p_{X|Y}(x|y) \equiv \mathbf{P}(X = x|Y = y)$ reads: The probability of $X = x$, given $Y = y$.

$$P(Y = y) = \sum_{i=1}^{n} P(Y = y | X = x_i) \cdot P(X = x_i).$$

Example 3.3.4 (How to avoid running into a bear?). Heuristically, one can think about conditional probabilities as probabilities obtained under additional constraints. Think here about the probability of your running into a bear during a hike. Given that you are hiking in the Cleveland Metroparks, the probability of the event may be only 0.0001; in Yellowstone the similar conditional probability may be as high as 0.75. Now assume you participate, with 51 of your classmates, in a raffle and the prize is a trip to Yellowstone; the consolation prize is a group hike in the Metroparks. The total probability of your running into a bear would then be $0.0001 \cdot (51/52) + 0.75 \cdot (1/52) \approx 0.015$.

One of the corollaries of the total probability formula is the celebrated *Bayes' formula for reverse conditional probabilities*, which, loosely speaking, computes the conditional probability of X, given Y, in terms of the conditional probabilities of Y, given X:

$$P(X = x_i | Y = y) = \frac{P(Y = y | X = x_i) \cdot P(X = x_i)}{\sum_{i=1}^{n} P(Y = y | X = x_i) \cdot P(X = x_i)}.$$

Indeed,

$$P(X = x_i | Y = y) = \frac{P(X = x_i, Y = y)}{P(Y = y)} \cdot \frac{P(X = x_i)}{P(X = x_i)}$$
$$= \frac{P(Y = y | X = x_i) \cdot P(X = x_i)}{P(Y = y)},$$

and an application of the total probability formula immediately gives the final result.

Example 3.3.5 (Transmission of a binary signal in the presence of random errors). A channel transmits the binary symbols 0 and 1 with random errors. The probability that the symbols 0 and 1 appear at the input of the channel is, respectively, 0.45 and 0.55. Because of transmission errors, if the symbol 0 appears at the input, then the probability of it being received as 0 at the output is 0.95. The analogous conditional probability is 0.9 for the symbol 1 to be received, given that it was transmitted. Our task is to find the reverse conditional probability that the symbol 1 was transmitted given that 1 was received.

The random vector here is (X, Y), where X is the input signal and Y is the output signal. The problem's description contains the following information:

$$P(X = 0) = 0.45, \qquad P(X = 1) = 0.55,$$

and

$$P(Y = 0 | X = 0) = 0.95, \qquad P(Y = 1 | X = 1) = 0.9,$$

so that

$$P(Y = 1 | X = 0) = 0.05, \qquad P(Y = 0 | X = 1) = 0.1.$$

We are seeking $P(X = 1|Y = 1)$, and Bayes' formula gives the answer:

$$P(X = 1|Y = 1) = \frac{P(Y = 1|X = 1) \cdot P(X = 1)}{P(Y = 1|X = 0) \cdot P(X = 0) + P(Y = 1|X = 1) \cdot P(X = 1)}$$

$$= \frac{0.9 \cdot 0.55}{0.05 \cdot 0.45 + 0.9 \cdot 0.55} \approx 0.9565.$$

Statistical independence. The components X and Y of a random vector $X = (X, Y)$ are said to be *statistically independent* if the conditional probabilities of X given Y are independent of Y, and vice versa. In the discrete case, this means that, for all x and y,

$$P(X = x|Y = y) = P(X = x),$$

which is equivalent to the statement that the joint p.d.f. is the product of the marginal p.d.f.s. Indeed, the above independence assumption and the formula defining the conditional probabilities yield

$$P(X = x, Y = y) = P_{(X,Y)}(x, y)$$
$$= P_X(x) \cdot P_Y(y) = P(X = x) \cdot P(Y = y). \quad (3.3.11)$$

In the continuous case the analogous definition of the independence of X and Y can be stated via the multiplicative formula for the joint p.d.f.:

$$f_{(X,Y)}(x, y) = f_X(x) \cdot f_Y(y). \qquad (3.3.12)$$

Note that both the 2D Bernoulli distribution of Example 3.3.1 and the 2D Gaussian distribution of Example 3.3.2 have statistically independent components X and Y. Also, components of the random vector (X, Y) in Example 3.3.3 are independent since the table was actually obtained by multiplying the marginal probabilities in the corresponding rows and columns. However, the components W and Z of the random vector (W, Z) in Example 3.3.3 are not statistically independent. To see this, it is sufficient to observe that

$$P(W = 3, Z = 2) = 0,$$

but

$$P(W = 3) \cdot P(Z = 2) = 3/12 \cdot 6/12 = 18/144 \neq 0.$$

Moments of random vectors and correlations. If a random quantity Z is a function of a random vector (X, Y), say,

$$Z = g(X, Y),$$

then, as in Sect. 3.2, we can calculate the mean of Z using the joint p.d.f. Indeed,

$$EZ = \sum_x \sum_y g(x, y) p_{(X,Y)}(x, y) \qquad (3.3.13)$$

in the discrete case, and

$$EZ = \int_{-\infty}^{\infty} \int_{-\infty}^{\infty} g(x, y) f_{(X,Y)}(x, y)\, dx\, dy \qquad (3.3.14)$$

in the continuous case.

A mixed central second-order moment corresponding to function $g(x, y) = (x - \mu_X)(y - \mu_Y)$ will play a pivotal role in the analysis of random signals. The number

$$\text{Cov}\,(X, Y) = \mathbf{E}\Big[(X - \mu_X)(Y - \mu_Y)\Big] = \mathbf{E}(XY) - \mathbf{E}(X)\mathbf{E}(Y), \qquad (3.3.15)$$

is called the *covariance* of X and Y. Obviously, the covariance of X and X is just the variance of X:

$$\text{Cov}\,(X, X) = \mathbf{E}\Big[(X - \mu_X)(X - \mu_Y)\Big] = \text{Var}\,(X). \qquad (3.3.16)$$

In the case when the expectations of X and Y are zero,

$$\text{Cov}\,(X, X) = \mathbf{E}(X \cdot Y). \qquad (3.3.17)$$

By the Cauchy–Schwartz inequality,[14]

$$|\text{Cov}\,(X, Y)| \le \text{Std}(X) \cdot \text{Std}(Y). \qquad (3.3.18)$$

This suggests the introduction of yet another parameter for a 2D random vector which is called the *correlation coefficient* of X and Y:

$$\text{Cor}\,(X, Y) \equiv \rho_{X,Y} = \frac{\text{Cov}\,(X, Y)}{\text{Std}(X) \cdot \text{Std}(Y)}. \qquad (3.3.19)$$

In view of (3.3.18), the correlation coefficient is always contained between -1 and $+1$:

$$-1 \le \rho_{X,Y} \le 1, \qquad (3.3.20)$$

---

[14] Recall that if $\mathbf{a} = (a_1, \ldots, a_d)$ and $\mathbf{b} = (b_1, \ldots, b_d)$ are two d-dimensional vectors, then the Cauchy–Schwartz inequality says that the absolute value of their scalar (dot) product is not larger than the product of their norms (magnitudes), i.e., $|\langle \mathbf{a}, \mathbf{b} \rangle| \le \|\mathbf{a}\| \cdot \|\mathbf{b}\|$, where $\langle \mathbf{a}, \mathbf{b} \rangle = a_1 b_1 + \cdots + a_d b_d$, and $\|\mathbf{a}\|^2 = a_1^2 + \cdots + a_d^2$; see Sect. 3.7.

and, in view of (3.3.17), if the random components X and Y are linearly dependent, that is, $Y = \alpha X$, then the correlation coefficient takes its extreme values

$$\rho_{X,\alpha X} = \pm 1, \tag{3.3.21}$$

depending on whether α is positive or negative. In those cases we say that the random quantities X and Y are perfectly (positively, or negatively) correlated. If $\rho_{X,Y} = 0$, then the random quantities X and Y are said to be *uncorrelated*.

The opposite case is that of statistically independent random quantities X and Y. Then, because of the multiplicative property $f_{(X,Y)}(x, y) = f_X(x) f_Y(y)$ (3.3.11) and (3.3.12) of the joint p.d.f., we always have

$$\mathbf{E}(XY) = \int\int xy f_X(x) f_Y(y)\, dx\, dy = \mathbf{E}X \cdot \mathbf{E}Y, \tag{3.3.22}$$

so that

$$\mathrm{Cov}\,(X, Y) = \mathbf{E}(XY) - \mathbf{E}X \cdot \mathbf{E}Y = 0, \tag{3.3.23}$$

and the correlation coefficient $\rho_{X,Y} = 0$. Thus *independent random quantities are always uncorrelated*. In this context, the correlation coefficient $\rho_{X,Y}$ is often considered as a measure of the "independence" of the random quantities X and Y; more appropriately, it should be interpreted as a measure of the "linear association" of the random quantities X and Y.

Remark 3.3.1 (Uncorrelated random quantities need not be independent). Although statistically independent random quantities are always uncorrelated, the reverse implication is not true in general. Indeed, consider an example of a 2D random vector (X, Y) with values uniformly distributed inside the unit circle. Obviously, because of the symmetry, $\mathbf{E}X = \mathbf{E}Y = 0$, and the covariance (calculated in polar coordinates) is

$$\mathrm{Cov}(X, Y) = \int_{-1}^{1}\int_{-\sqrt{1-x^2}}^{+\sqrt{1-x^2}} xy \frac{dy\, dx}{\pi} = \int_{0}^{2\pi}\int_{0}^{1} r^3 \cos\theta \sin\theta \, \frac{dr\, d\theta}{\pi} = 0.$$

So X and Y are uncorrelated. But they are not independent, because, for example,

$$\mathbf{P}(\sqrt{2}/2 < X < 1, \sqrt{2}/2 < Y < 1) \neq \mathbf{P}(\sqrt{2}/2 < X < 1) \cdot \mathbf{P}(\sqrt{2}/2 < Y < 1).$$

Indeed, the left-hand side is zero since the square $\{(x, y) : \sqrt{2}/2 < x < 1, \sqrt{2}/2 < y < 1\}$ lies outside the unit circle, but the right-hand side is positive since

$$\mathbf{P}(\sqrt{2}/2 < X < 1) = \int_{\sqrt{2}/2}^{1}\int_{-\sqrt{1-x^2}}^{+\sqrt{1-x^2}} \frac{dy\, dx}{\pi} = \mathbf{P}(\sqrt{2}/2 < Y < 1) > 0;$$

each of the above probabilities is simply the (normalized) area of the sliver of the unit disk to the right of the vertical line $x = \sqrt{2}/2$. However, in certain special cases the reverse implication is true: Gaussian random quantities are independent if and only if they are uncorrelated; see Chap. 8.

Example 3.3.6 (A discrete 2D distribution with nontrivial correlation). Consider the random vector (W, Z) from Example 3.3.3. The expectations of the components are

$$\mathbf{E}W = 1(5/12) + 2(4/12) + 3(3/12) = 11/6,$$
$$\mathbf{E}Z = 1(6/12) + 2(6/12) = 3/2.$$

The variances are

$$\mathrm{Var}(W) = (1 - 11/6)^2(5/12) + (2 - 11/6)^2(4/12)$$
$$+(3 - 11/6)^2(3/12) = 23/36,$$
$$\mathrm{Var}(Z) = (1 - 3/2)^2(6/12) + (2 - 3/2)^2(6/12) = 1/4.$$

The expectation of the product is

$$\mathbf{E}(WZ) = (1 \cdot 1)(1/12) + (2 \cdot 1)(2/12) + (3 \cdot 1)(3/12)$$
$$+ (1 \cdot 2)(4/12) + (2 \cdot 2)(2/12) + (3 \cdot 2)0 = 5/2.$$

Thus the covariance is

$$\mathrm{Cov}(W, Z) = \mathbf{E}(WZ) - \mathbf{E}(W)\mathbf{E}(Z) = 5/2 - (11/6)(3/2) = -1/4,$$

and, finally, the correlation coefficient of W and Z is

$$\mathrm{Cor}(W, Z) = \frac{\mathrm{Cov}\,(W, Z)}{\mathrm{Std}(W) \cdot \mathrm{Std}(Z)} = \frac{-1/4}{\sqrt{23/36} \cdot \sqrt{1/4}} = -\sqrt{3/23} \approx -0.361.$$

Example 3.3.7 (A continuous 2D distribution with nontrivial correlation). A random vector (X, Y) has a continuous joint p.d.f. of the form

$$f_{(X,Y)}(x, y) = \begin{cases} C(1 - (x + y)), & \text{for } x, y \geq 0, x + y \leq 1; \\ 0, & \text{elsewhere.} \end{cases}$$

The constant C can be determined from the normalization condition,

$$\int_0^1 \int_0^{1-x} C(1 - (x + y))\, dy\, dx = 1,$$

which gives $C = 6$. The plot of the surface representing this density is given in Fig. 3.3.2.

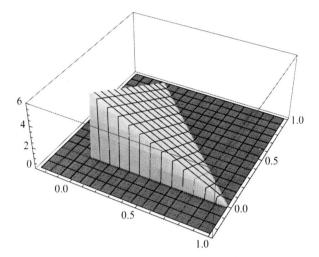

Fig. 3.3.2 The plot of the surface representing the joint p.d.f. from Example 3.3.7

Fig. 3.3.3 The marginal
density $F_X(x)$ of the X
component of the random
vector from Example 3.3.7

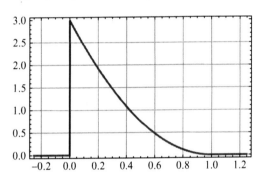

The marginal density of the component X is

$$f_X(x) = \int_0^{1-x} 6(1 - (x + y)) \, dy = 3(1 - x)^2,$$

for $0 < x < 1$. It is equal to 0 elsewhere, and its plot is pictured in Fig. 3.3.3.
The expectations of X and Y are easily evaluated using the marginal p.d.f.:

$$EX = EY = \int_0^1 x \cdot 3(1 - x)^2 \, dx = \frac{1}{4}.$$

Similarly, the variances are

$$\sigma^2(X) = \sigma^2(Y) = \int_0^1 (x - 1/4)^2 \cdot 3(1 - x)^2 \, dx = \frac{3}{80}.$$

Finally, the covariance is

$$\text{Cov}\,(X,Y) = \int_0^1 \int_0^{1-x} (x-1/4)(y-1/4)\cdot 6(1-(x+y))\,dy\,dx = -\frac{1}{80}.$$

So the random components X and Y are not independent; they are negatively correlated. The correlation coefficient itself is now easily evaluated to be

$$\rho_{X,Y} = \frac{-1/80}{3/80} = -\frac{1}{3}.$$

3.4 The Least-Squares Fit, Linear Regression

The roles of the covariance and the correlation coefficient will become better understood in the context of the following *least-squares regression* problem. Consider a sample,

$$(x_1, y_1), (x_2, y_2), \ldots, (x_N, y_N),$$

of N 2D vectors. Its representation in the (x, y)-plane is called the *scatter plot* of the sample; see, for example, Fig. 3.4.1. Our goal is to find a line,

$$y = ax + b,$$

which would provide the best approximation to the scatterplot in the sense of minimizing the sum of the squares of the errors of the approximation measured in the vertical direction. To be more precise, the error of the approximation for the ith sample point is expressed by the formula

$$\epsilon_i(a,b) = |y_i - (ax_i - b)|, \qquad i = 1, 2, \ldots, N,$$

and the sum of the squares of the errors,

$$\sum_{i=1}^{N} \epsilon_i^2(a,b) = \sum_{i=1}^{N} (y_i - (ax_i - b))^2,$$

is a nice, differentiable function of two variables a and b. We can find its minimum by taking partial derivatives with respect to a and b and equating them to 0[15]:

---

[15] This explains why we consider quadratic errors rather than the straight absolute errors; in the latter case the calculus tools would not work so well.

$$\frac{\partial}{\partial a} \sum_{i=1}^{N} \epsilon_i^2(a,b) = -2 \sum_{i=1}^{N} (y_i - (ax_i + b))x_i = 0,$$

$$\frac{\partial}{\partial b} \sum_{i=1}^{N} \epsilon_i^2(a,b) = -2 \sum_{i=1}^{N} (y_i - (ax_i + b)) = 0.$$

These two equations, sometimes called the *normal equations*, are linear in a and b and can be easily solved by the substitution method. To make the next step more transparent, we will introduce the following simplified notation for different sample means (think here about the means of random quantities with N possible values, with each value assigned probability $1/N$). The x and y components of the above data will be treated as ND vectors and denoted

$$x = (x_1, \ldots, x_N), \quad y = (y_1, \ldots, y_N).$$

Various sample means will be denoted as follows:

$$\overline{x} = \frac{1}{N} \sum_{i=1}^{M} x_i, \quad \overline{y} = \frac{1}{N} \sum_{i=1}^{M} y_i,$$

$$\overline{x^2} = \frac{1}{N} \sum_{i=1}^{M} x_i^2, \quad \overline{y^2} = \frac{1}{N} \sum_{i=1}^{M} y_i^2,$$

$$\overline{xy} = \frac{1}{N} \sum_{i=1}^{M} x_i y_i.$$

Now, the normal equations for a and b can be written in the form

$$a\overline{x} + b - \overline{y} = 0 \quad \text{and} \quad a\overline{x^2} + b\overline{x} - \overline{xy} = 0,$$

which can be immediately solved to give

$$b = \overline{y} - a\overline{x}, \quad a = \frac{\overline{xy} - \overline{x} \cdot \overline{y}}{\overline{x^2} - (\overline{x})^2}.$$

The first of the above two equations indicates that the point with coordinates formed by the sample means $\overline{x}$ and $\overline{y}$ is located on the regression line. To better see the meaning of the second equation, observe that

$$\overline{xy} - \overline{x} \cdot \overline{y} = \frac{1}{N} \sum_{i=1}^{N} (x_i - \overline{x})(y_i - \overline{y}) = \text{Cov}(x, y)$$

is just the sample covariance of the x- and y-coordinates of 2D data, and that

$$\overline{x^2} - (\overline{x})^2 = \mathrm{Var}(x), \qquad \overline{y^2} - (\overline{y})^2 = \mathrm{Var}(y).$$

Thus the equation $y = ax + b$ of the regression line now becomes

$$y = \frac{\mathrm{Cov}(x,y)}{\mathrm{Var}(x)}x + \left(\overline{y} - \frac{\mathrm{Cov}(x,y)}{\mathrm{Var}(x)}\right)\overline{x},$$

and can finally be rewritten in a more elegant and symmetric form,

$$\frac{y - \overline{y}}{\mathrm{Std}(y)} = \rho_{x,y} \cdot \frac{x - \overline{x}}{\mathrm{Std}(x)}, \tag{3.4.1}$$

where

$$\rho_{x,y} = \frac{\mathrm{Cov}(x,y)}{\mathrm{Std}(x)\sqrt{\mathrm{Std}(y)}}$$

is the sample correlation coefficient; the standard deviation Std, as usual, denotes the square root of the variance Var. The significance of the form of the regression equation (3.4.1) is now clear: $\rho_{x,y}$ is the slope of the regression line but only after the x- and y-coordinates were standardized [see (3.2.11)]; that is, they were centered by the means $\overline{x}$ and $\overline{y}$, and rescaled by the standard deviations $\mathrm{Std}(x)$ and $\mathrm{Std}(y)$, respectively.

Example 3.4.1. Consider a 2D vector sample of size 10:

x	y
1.05983	1.10539
2.07758	3.36697
3.28160	3.22934
4.13003	6.91638
5.28022	7.65665
6.38872	6.78509
7.11893	8.11736
8.04133	9.94112
9.23407	9.55498
10.3814	10.8697

The coefficients are $a = 0.9934$ and $b = 1.0925$, so that the equation of the regression line is

$$y = 0.9934 \cdot x + 1.0925,$$

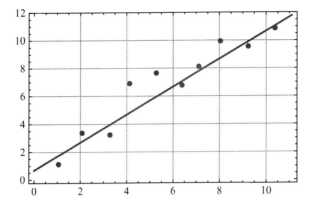

Fig. 3.4.1 The scatterplot and the least-squares fit regression line for data from Example 3.4.1

and the correlation coefficient,

$$\rho_{x,y} = 0.9503,$$

turns out to be relatively close to 1, indicating a strong positive "linear association" between the x- and y-data. The scatterplot of these data as well as the plot of the regression line (best linear fit) are shown in Fig. 3.4.1.

3.5 The Law of Large Numbers and the Stability of Fluctuations Law

One of the fundamental theorems of statistics, called the *law of large numbers* (*LLN*), says that if $X_1, X_2, \ldots, X_n$, are independent random quantities with identical probability distributions, and finite identical expectations $EX_i = \mu_X$, then, as $n \to \infty$, the averages converge to that expectation, i.e.,

$$\bar{X}_n \equiv \frac{X_1 + X_2 + \cdots + X_n}{n} \longrightarrow \mu_X, \quad \text{as} \quad n \to \infty. \quad (3.5.1)$$

Of course, the immediate issue is what we mean here by the convergence of random variables $\bar{X}_n$. For the purpose of these lectures, the convergence of $\bar{X}_n$ to μ_X will mean that the standard deviation of the fluctuations of the averages $\bar{X}_n$ around the mean μ_X, that is, the differences $\bar{X}_n - \mu_X$, converges to zero as $n \to \infty$. More formally,

$$\lim_{n \to \infty} \text{Std}(\bar{X}_n - \mu_X) = 0. \quad (3.5.2)$$

The statement (3.5.2) can be easily verified if we observe first that:

(a) For any random vector (X, Y) with finite absolute first moments of the components, the expectation

$$\mathbf{E}(X + Y) = \mathbf{E}(X) + \mathbf{E}(Y).$$ (3.5.3a)

Indeed, taking $g(x, y) = x + y$ in formulas (3.3.14) and (3.3.15) defining expectations of functions of random vectors so that (say, in the continuous case)

$$\mathbf{E}(X + Y) = \int_{-\infty}^{\infty} \int_{-\infty}^{\infty} (x + y) f_{(X,Y)}(x, y) \, dx \, dy$$

$$= \int_{-\infty}^{\infty} x \left(\int_{-\infty}^{\infty} f_{(X,Y)}(x, y) \, dy \right) dx$$

$$+ \int_{-\infty}^{\infty} y \left(\int_{-\infty}^{\infty} f_{(X,Y)}(x, y) \, dx \right) dy$$

$$= \int_{-\infty}^{\infty} x f_X(x) \, dx + \int_{-\infty}^{\infty} y f_Y(y) \, dy = \mathbf{E}(X) + \mathbf{E}(Y),$$

in view of formula (3.3.9) for the marginal p.d.f. of a component of a random vector.[16]

(b) For any random (X, Y) with *independent* components with finite variances, the variance

$$\mathrm{Var}(X + Y) = \mathrm{Var}(X) + \mathrm{Var}(Y).$$ (3.5.3b)

This follows immediately from the multiplicative property (3.3.22) of the expectations of independent random variables; see Sect. 3.3.

Now, if X and Y are independent, then $X - \mu_X$ and $Y - \mu_Y$ are also independent, so that, utilizing (3.5.3a) and (3.5.3b),

$$\mathrm{Var}(X + Y) = \mathbf{E}((X - \mu_X) + (Y - \mu_Y))^2$$
$$= \mathbf{E}(X - \mu_X)^2 + 2\mathbf{E}(X - \mu_X)\mathbf{E}(Y - \mu_Y) + \mathbf{E}(Y - \mu_Y)^2$$
$$= \mathrm{Var}(X) + \mathrm{Var}(Y),$$

because $\mathbf{E}(X - \mu_X) = \mathbf{E}(Y - \mu_Y) = 0$. Hence,

$$\mathrm{Var}(\bar{X}_n - \mu_X) = \mathrm{Var}\left(\frac{X_1 - \mu_X}{n} + \cdots + \frac{X_n - \mu_X}{n}\right) = \frac{\mathrm{Var}(X)}{n},$$ (3.5.4)

---

[16] Note how the knowledge of the joint probability distribution of the random vector (X, Y), and also of $(X_1, X_2, \ldots, X_n)$, is what permits us to study the sums $X + Y$ and $X_1 + X_2 + \cdots + X_n$ as real-valued random quantities with well-defined probability distributions; see Remarks 3.2.1 and 3.5.1.

which obviously approaches 0 as $n \to \infty$. Thus the law of large numbers (3.5.1), also often called the *law of averages*, is verified, at least in the situation when random quantities X_i have well-defined finite variances.[17]

More subtle information about the averages is provided by the following *stability of fluctuations law*, usually called the *central limit theorem (CLT)* in the mathematical and statistical literature. It states that as the averages $\bar{X}_n$ fluctuate around the expectation μ_X, the fluctuations, if viewed under a "magnifying glass," turn out to follow, asymptotically as $n \to \infty$, a Gaussian or normal probability distribution. More precisely, the c.d.f. of the standardized [see (3.2.13)] random fluctuations of the averages $\bar{X}_n$ around the mean μ_X,

$$Z_n = \frac{\sqrt{n}}{\text{Std}(X)} \cdot (\bar{X}_n - \mu_X), \tag{3.5.5}$$

converges to the standard $N(0, 1)$ Gaussian c.d.f.; that is,

$$\lim_{n \to \infty} \mathbf{P}(Z_n \leq z) = \Phi(z) \equiv \int_{-\infty}^{z} \phi(x)\, dx, \tag{3.5.6}$$

where the density

$$\phi(z) = \frac{1}{\sqrt{2\pi}} e^{-z^2/2} \tag{3.5.7}$$

is that of the standard $N(0, 1)$ Gaussian random quantity. The important assumption of the central limit theorem is that the common variance of X_is is finite.

Summarizing the above discussion, the central limit theorem can be loosely rephrased as follows:

Standardized random fluctuations of averages of independent and identically distributed random quantities around their common expected value have a limiting standard Gaussian cumulative distribution function.

Remark 3.5.1 (Probability distribution of a sum of independent random quantities.). It can be immediately verified that all of the Z_ns in (3.5.5) have mean zero and variance one [see (3.2.13) and (3.5.3)], but the proof of the convergence to a Gaussian limit is more delicate. Without going into the details (for a sketch of the full proof, see Sect. 3.7), it is clear that the proof has to rely on the determination of the probability distribution of the sum $Z = X + Y$ of two (or more) independent random quantities X and Y. In the case of continuous random quantities (for the derivation in case of discrete random quantities, see Sect. 3.7), it turns out that the p.d.f. of $Z = X + Y$ is the convolution of the p.d.f.s of X and Y. Indeed, in view of the independence of X and Y, the c.d.f. of Z, for an arbitrary but fixed z, is equal to

$$F_Z(z) = \mathbf{P}(Z \leq z) = \mathbf{P}(X + Y \leq z) = \iint_{\{(x,y): x+y \leq z\}} f_{(X,Y)}(x, y)\, dx\, dy$$

$$= \int_{-\infty}^{\infty} \int_{-\infty}^{z-y} f_X(x) f_Y(y)\, dx\, dy = \int_{-\infty}^{\infty} \left(\int_{-\infty}^{z-y} f_X(x)\, dx \right) f_Y(y)\, dy$$

---

[17] Observe that not all random quantities have well-defined, finite variances; see Problem 3.7.28.

$$= \int_{-\infty}^{\infty} \left(\int_{-\infty}^{z} f_X(u - y) \, du \right) f_Y(y) \, dy$$

$$= \int_{-\infty}^{z} \left(\int_{-\infty}^{\infty} f_X(u - y) f_Y(y) \, dy \right) du,$$

after a change of variables, $x = u - y$, and then a change of the order of integration. Consequently, the p.d.f.

$$f_Z(z) = f_{X+Y}(z) = \int_{-\infty}^{\infty} f_X(z - y) f_Y(y) \, dy = (f_X * f_Y)(z). \qquad (3.5.8)$$

As we have seen in Chap. 2, convolution can be a fairly complex operation even in the case of relatively simple $f_X(x)$ and $f_Y(y)$. Moreover, the distribution of $X_1 + \cdots + X_n$ in (3.5.1) is an n-fold convolution of the p.d.f. $f_X(x)$, and the n is growing to infinity. So dealing directly with the p.d.f. of the average $\bar{X}_n$, $n \to \infty$, seems to be a hopeless task. However, in view of Chap. 2, it is obvious that the whole problem would be greatly simplified if, instead of dealing with p.d.f.s, one could employ their Fourier transforms; the convolution is replaced in the frequency domain by simple pointwise products. This idea is implemented in the sketch of the proof suggested in Problem 3.7.24.

3.6 Estimators of Parameters and Their Accuracy; Confidence Intervals

The law of large numbers can be reinterpreted as follows: If $X_1, X_2, \ldots, X_n$ are independent and identically distributed random quantities representing repeated sampling from a certain probability distribution $F_X(x)$, then, as n increases, the sample means $\bar{X}_n, n = 1, 2, \ldots$, become better and better estimators of the expectation of that distribution. In statistical terminology the law of large numbers (3.5.1) says that $\bar{X}_n$ is a *consistent estimator* for parameter μ_X.

The central limit theorem (3.5.5) and (3.5.6) permits us to say what is, the error of approximation of the theoretical mean μ_X by the sample mean $\bar{X}_n$ or, in other words, to establish the accuracy of the above estimation. Indeed, for a given sample of size n, the CLT says that the difference between the parameter μ_X and its estimator, the sample mean $\bar{X}_n$, is, after normalization by $\sqrt{n}/\mathrm{Std}(X)$, approximately $N(0, 1)$-distributed, so that, for large n,

$$\mathbf{P}\left(-\epsilon \frac{\mathrm{Std}(X)}{\sqrt{n}} \leq \bar{X}_n - \mu_X \leq \epsilon \frac{\mathrm{Std}(X)}{\sqrt{n}} \right) \approx \Phi(\epsilon) - \Phi(-\epsilon) = 2\Phi(\epsilon) - 1, \ (3.6.1)$$

where $\Phi(z)$ is the c.d.f. of the standard Gaussian ($N(0, 1)$) random quantity tabulated in Table 3.6.1.

Table 3.6.1 Gaussian $N(0, 1)$ c.d.f.: $\Phi(z) = (2\pi)^{-1/2} \int_{-\infty}^{z} e^{-x^2/2}\, dx$

z	0.00	0.01	0.02	0.03	0.04	0.05	0.06	0.07	0.08	0.09
0.0	0.5000	0.5040	0.5080	0.5120	0.5160	0.5199	0.5239	0.5279	0.5319	0.5359
0.1	0.5395	0.5438	0.5478	0.5517	0.5557	0.5596	0.5636	0.5675	0.5714	0.5753
0.2	0.5793	0.5832	0.5871	0.5910	0.5948	0.5987	0.6026	0.6064	0.6103	0.6141
0.3	0.6179	0.6217	0.6255	0.6296	0.6331	0.6366	0.6406	0.6443	0.6480	0.6517
0.4	0.6554	0.6591	0.6628	0.6664	0.6700	0.6736	0.6772	0.6808	0.6884	0.6879
0.5	0.6915	0.6956	0.6985	0.7019	0.7054	0.7088	0.7123	0.7157	0.7190	0.7224
0.6	0.7257	0.7291	0.7324	0.7857	0.7389	0.7422	0.7454	0.7486	0.7517	0.7549
0.7	0.7580	0.7611	0.7642	0.7673	0.7704	0.7734	0.7764	0.7794	0.7823	0.7852
0.8	0.7881	0.7910	0.7939	0.7967	0.7995	0.8023	0.8051	0.8075	0.8106	0.8133
0.9	0.8195	0.8186	0.8212	0.8238	0.8264	0.8289	0.8315	0.8340	0.8365	0.8389
1.0	0.8413	0.8438	0.8461	0.8485	0.8503	0.8531	0.8554	0.8577	0.8599	0.8621
1.1	0.8613	0.8665	0.8686	0.8708	0.8729	0.8749	0.8770	0.8796	0.8810	0.8830
1.2	0.8849	0.8869	0.8888	0.8907	0.8925	0.8944	0.8962	0.8980	0.8977	0.9015
1.3	0.9032	0.9049	0.9066	0.9082	0.9099	0.9115	0.9131	0.9147	0.9162	0.9177
1.4	0.9192	0.9207	0.9222	0.9236	0.9251	0.9265	0.9279	0.9292	0.9306	0.9319
1.5	0.9332	0.9345	0.9359	0.9370	0.9382	0.9309	0.9404	0.9418	0.9429	0.9441
1.6	0.9452	0.9463	0.9474	0.9484	0.9495	0.9505	0.9515	0.9525	0.9535	0.9545
1.7	0.9554	0.9564	0.9573	0.9582	0.9591	0.9599	0.9606	0.9616	0.9625	0.9633
1.8	0.9641	0.9649	0.9656	0.9664	0.9671	0.9678	0.9666	0.9693	0.9699	0.9706
1.9	0.9713	0.9719	0.9726	0.9732	0.9738	0.9744	0.9750	0.9756	0.9761	0.9767
2.0	0.9773	0.9778	0.9783	0.9788	0.9793	0.9798	0.9803	0.9808	0.9812	0.9817
2.1	0.9821	0.9826	0.9830	0.9834	0.9838	0.9842	0.9846	0.9850	0.9854	0.9857
2.2	0.9891	0.9861	0.9868	0.9871	0.9875	0.9878	0.9881	0.9884	0.9887	0.9890
2.3	0.9893	0.9896	0.9868	0.9871	0.9875	0.9878	0.9881	0.9884	0.9887	0.9890
2.4	0.9918	0.9820	0.9922	0.9925	0.9927	0.9929	0.9931	0.9932	0.9934	0.9936
2.5	0.9938	0.9940	0.9941	0.9943	0.9945	0.9946	0.9948	0.9949	0.9951	0.9952
2.6	0.9953	0.9955	0.9956	0.9957	0.9959	0.9960	0.9961	0.9962	0.9963	0.9964
2.7	0.9965	0.9966	0.9967	0.9968	0.9969	0.9970	0.9971	0.9972	0.9973	0.9974
2.8	0.9974	0.9975	0.9976	0.9977	0.9977	0.9978	0.9979	0.9979	0.9980	0.9981
2.9	0.9981	0.9982	0.9983	0.9983	0.9984	0.9984	0.9985	0.9985	0.9986	0.9986
3.0	0.9987	0.9987	0.9987	0.9988	0.9988	0.9989	0.9989	0.9989	0.9990	0.9990
3.1	0.9990	0.9991	0.9991	0.9991	0.9992	0.9992	0.9992	0.9992	0.9993	0.9993
3.2	0.9993	0.9993	0.9994	0.9994	0.9994	0.9994	0.9994	0.9995	0.9995	0.9995
3.3	0.9995	0.9995	0.9996	0.9996	0.9996	0.9996	0.9996	0.9996	0.9996	0.9997
3.4	0.9997	0.9997	0.9997	0.9997	0.9997	0.9997	0.9997	0.9997	0.9997	0.9998

If X itself has a Gaussian p.d.f., the above approximate equality becomes exact for all n. This follows from the fact that the sum of two independent Gaussian random quantities is again a Gaussian random quantity, obviously with the mean and variance being the sums of means and variances, respectively, of the corresponding random summands; see Sect. 3.7.

The contents of formula (3.6.1) can be rephrased as follows: The true value of *parameter μ_X is contained in the random interval*

$$\left(\bar{X}_n - \epsilon \frac{\mathrm{Std}(X)}{\sqrt{n}}, \bar{X}_n + \epsilon \frac{\mathrm{Std}(X)}{\sqrt{n}} \right)$$

with probability

$$C = C(\epsilon) = 2\Phi(\epsilon) - 1.$$

The above random interval is called the *confidence interval*, and the probability $C = C(\epsilon)$ is called its *confidence level*. The above statement is sometimes abbreviated by writing

$$\mu_X = \bar{X}_n \pm \epsilon \frac{\mathrm{Std}(X)}{\sqrt{n}}$$

at the confidence level C. Note that it is the center of the above random interval that is random; its length is not random unless $\mathrm{Std}(X)$ itself has to be estimated from the sample.

Example 3.6.1 (A 95% confidence interval for μ_X with known $\mathrm{Std}(X)$). Sixteen independently repeated measurements of a random quantity X were conducted, resulting in $\bar{X}_{16} = 2.56$. Suppose that we know that $\mathrm{Std}(X) = 0.12$. To find the 95% confidence interval for μ_X using (3.6.1), we need to find ϵ such that $2\Phi(\epsilon) - 1 = 0.95$; i.e., $\Phi(\epsilon) = 0.975$. From Table 3.6.1 of the Gaussian $N(0, 1)$ c.d.f., we have $\epsilon = 1.96$. Thus, at the 95% confidence level,

$$2.56 - 1.96\frac{0.12}{\sqrt{16}} \le \mu_X \le 2.56 + 1.96\frac{0.12}{\sqrt{16}};$$

that is,

$$\mu_X = 2.56 \pm 0.059$$

at the 95% confidence level. The above approximate confidence interval is exact if X has a Gaussian distribution.

Remark 3.6.1 (Error of the Gaussian approximation in the CLT). To be honest, we left open the essential, but delicate, question of how good is the approximate equality in the basic formula (3.6.1), or, equivalently, the question of how precise is the estimation of the error in the central limit theorem (3.5.6), which, by itself, only says that the difference

$$\mathbf{P}(Z_n \le z) - \Phi(z) \to 0, \qquad \text{as} \qquad n \to \infty,$$

where

$$Z_n = \frac{(X_1 + \cdots + X_n) - n\mu_X}{\sqrt{n} \cdot \mathrm{Std}(X)}$$

are standardized sums $X_1 + \cdots + X_n$. It turns out that the accuracy in the CLT is actually pretty good if the X_is have higher absolute moments finite. In particular, if the third central moment $m_3 = \mathbf{E}|X - \mu_X|^3 < \infty$ then, for all $-\infty < x < \infty$ and $n = 1, 2, \ldots,$

$$|\mathbf{P}(Z_n \leq z) - \Phi(z)| \leq \frac{k m_3}{\sqrt{n}(\mathrm{Std}(X))^3},$$

where k is a universal (independent of n and X) constant contained in the interval $(0.4097, 0.7975)$. Its exact value is not known.[18]

Of course, the procedure used in Example 3.6.1 requires advance knowledge of the standard deviation $\mathrm{Std}(X)$. If that parameter is unknown, then the obvious step is to try to estimate it from the sample $X_1, X_2, \ldots, X_n$ itself using the sample variance estimator

$$S_n^2 = \frac{1}{n-1} \sum_{i=1}^{n} (X_i - \bar{X})^2, \tag{3.6.2}$$

which is an unbiased estimator for $\mathrm{Var}(X)$; see, Problem 3.7.29.

But in this case, even if the X_is are Gaussian, the standardized random quantity

$$T = \frac{\sqrt{n}}{S_n}(\bar{X} - \mu_X) \tag{3.6.3}$$

is no longer $N(0, 1)$-distributed, so a simple construction of the confidence interval for μ_X using the Gaussian distribution is impossible.

However, in the narrower situation of a Gaussian random sample $X_1, X_2, \ldots, X_n$, it is known that the random quantity T has the p.d.f.

$$f_T(x; n-1) = \frac{\Gamma((n)/2)}{\sqrt{n\pi}\,\Gamma((n-1)/2)} \left(1 + \frac{x^2}{n-1}\right)^{-n/2}, \tag{3.6.4}$$

which, traditionally, is called Students t p.d.f. with $(n-1)$ degrees of freedom.[19] A sample of different Student t p.d.f.s is shown in Fig. 3.6.1.

The gamma function $\Gamma(\gamma)$ appearing in the definition of f_T is defined by the formula

$$\Gamma(\gamma) = \int_0^{\infty} x^{\gamma-1} e^{-x}\, dx, \qquad \gamma > 0. \tag{3.6.5}$$

It is worth noting that

$$\gamma \Gamma(\gamma) = \Gamma(\gamma + 1) \qquad \text{and} \qquad \Gamma(n) = (n-1)! \tag{3.6.6}$$

---

[18] This error estimate in the CLT is known as the Berry–Esseen theorem and its proof can be found, for example, in V. V. Petrov's monograph *Sums of Independent Random Variables*, Springer, New York, 1975.

[19] See, for example, M. Denker and W. A. Woyczyński, *Introductory Statistics and Random Phenomena: Uncertainty, Complexity and Chaotic Behavior in Engineering and Science*, Birkhäuser Boston, Cambridge, MA, 1998, for more details on the statistical issues discussed in this section.

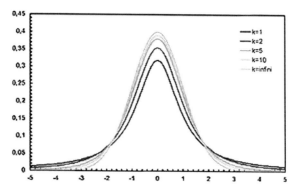

Fig. 3.6.1 Student's t p.d.f.s defined in (3.6.4), with (*bottom* to *top*) 1, 2, 5, 15, and ∞ degrees of freedom (from http://en.wikipedia.org/wiki/Student's t-distribution)

if n is a positive integer. So the gamma function is an interesting extension of the concept of the factorial to noninteger numbers.

Thus, in the Gaussian case with unknown variance, the confidence interval for μ_X at the confidence level $C = (2F_T(\epsilon) - 1)$ is of the form

$$\left(\bar{X}_n - \epsilon \frac{S_n}{\sqrt{n}}, \ \bar{X}_n + \epsilon \frac{S_n}{\sqrt{n}} \right). \tag{3.6.7}$$

Since in practice the goal is often to construct confidence intervals at given confidence levels, instead of tabulating the Student's t c.d.f.s $F_T(t)$, it is convenient to tabulate the relevant probabilities via the tail quantile function $q(\alpha; n)$ defined by the equality

$$q(\alpha; n) = Q_T(1 - \alpha; n),$$

where the quantile function $Q_T(\alpha; n)$ (see Remark 3.1.2) is the inverse function to the c.d.f. $F_T(t)$ i.e.,

$$F_T(Q_T(\alpha; n)) = \alpha. \tag{3.6.8}$$

Thus the tail quantile $q(\alpha; n)$ is the number such that the probability that Student's t random quantity with n degrees of freedom if greater than α. Selected tail quantiles $q_T(\alpha; n)$ are provided in Table 3.6.2.

Using the tail quantiles $q_T(\alpha; n)$, we can now write the C-confidence level interval for μ_X simply in the form

$$\left(\bar{X}_n - q_T\left(\frac{1 - C}{2}, n - 1\right) \frac{S_n}{\sqrt{n}}, \ \bar{X}_n + q_T\left(\frac{1 - C}{2}, n - 1\right) \frac{S_n}{\sqrt{n}} \right). \tag{3.6.9}$$

The Student's t p.d.f.s are symmetric about zero and bell-shaped but flatter than the $N(0, 1)$ pd.f. (why?). For large values of N, say $n > 20$, they are practically indistinguishable from the standard Gaussian p.d.f. (why?; see Problem 3.7.18), and the latter can be used in the construction of confidence intervals even in the case of unknown variance.

Table 3.6.2 Tail quantiles $q_T(\alpha; n)$ of Student's t distribution

$n \backslash \alpha$	0.1000	0.0500	0.0250	0.0100	0.0050	0.0010	0.0005
1	3.078	6.314	12.706	31.821	63.657	318.317	636.61
2	1.886	2.920	4.303	6.965	9.925	22.326	31.598
3	1.638	2.353	3.182	4.451	5.841	10.213	12.924
4	1.533	2.132	2.776	3.747	4.604	7.173	8.610
5	1.476	2.015	2.571	3.365	4.032	5.893	8.610
6	1.440	1.943	2.447	3.143	3.707	5.208	5.959
7	1.415	1.895	2.365	2.998	3.500	4.785	5.408
8	1.397	1.860	2.306	2.896	3.355	4.501	5.041
9	1.383	1.833	2.262	2.821	3.250	4.297	4.781
10	1.372	1.813	2.228	2.764	3.169	4.144	4.587
11	1.364	1.796	2.201	2.718	3.106	4.025	4.437
12	1.356	1.782	2.179	2.681	3.055	3.930	4.318
13	1.350	1.771	2.160	2.650	3.012	3.852	4.221
14	1.345	1.761	2.145	2.624	2.977	3.787	4.141
15	1.341	1.753	2.131	2.602	2.947	3.733	4.073
16	1.337	1.746	2.120	2.584	2.921	3.686	4.015
17	1.333	1.740	2.110	2.567	2.898	3.646	3.965
18	1.330	1.734	2.101	2.553	2.879	3.610	3.992
19	1.328	1.729	2.093	2.540	2.861	3.579	3.883
20	1.325	1.725	2.086	2.528	2.845	3.552	3.849
21	1.323	1.721	2.080	2.518	2.831	3.527	3.819
22	1.321	1.717	2.074	2.508	2.819	3.505	3.792
23	1.320	1.714	2.069	2.500	2.807	3.485	3.768
24	1.318	1.711	2.064	2.492	2.797	3.467	3.745
25	1.316	1.708	2.059	2.485	2.787	3.450	3.725
26	1.315	1.706	2.056	2.479	2.779	3.435	3.707
27	1.314	1.703	2.052	2.473	2.771	3.421	3.690
28	1.312	1.701	2.049	2.467	2.763	3.408	3.674
29	1.311	1.699	2.045	2.462	2.756	3.396	3.659
30	1.311	1.697	2.042	2.457	2.750	3.385	3.646
40	1.303	1.684	2.021	2.423	2.704	3.307	3.551
60	1.296	1.671	2.000	2.390	2.660	3.232	3.460
120	1.289	1.658	1.980	2.358	2.617	3.160	3.373
∞	1.282	1.645	1.960	2.326	2.576	3.090	3.291

Example 3.6.2 (A 95% confidence interval for μ_X with unknown Std(X)). Sixteen independent measurements of a Gaussian random quantity X resulted in $\bar{X}_{16} = 2.56$ and $S_{16} = 0.12$. With the desired confidence level $C = 0.95$, Table 3.6.2 yields the tail quantile

$$q_T((1 - 0.95)/2; 15) = q_T(0.025; 15) = 2.13.$$

Hence the 95% confidence interval for the expectation μ_X is of the form

$$\left(2.56 - 2.13 \cdot \frac{0.12}{\sqrt{16}}, \ 2.56 + 2.13 \cdot \frac{0.12}{\sqrt{16}}\right),$$

or, in other words, $\mu_X = 2.56 \pm 0.064$ at the 95% confidence level. Observe that, not surprisingly, in the absence of the precise knowledge of the variance $\mathrm{Var}(X)$, which had to be replaced by the estimator S_{16}, this confidence interval is wider than that in Example 3.6.1 ($\mu_X = 2.56 \pm 0.059$ at the same 95% confidence level), where the value of the variance was assumed to be known exactly.

The final question in this section is: How good is the sample variance estimator S_n^2 introduced in (3.6.2)? Here again the answer is difficult for a general c.d.f. F_X. However, in the case of a Gaussian $N(\mu_X, \sigma_X^2)$ sample, one can prove that the nonnegative random quantity

$$\chi^2 = \frac{1}{\sigma_X^2} \sum_{i=1}^{n} (X_i - \bar{X}_n)^2 \tag{3.6.10}$$

has the p.d.f. of the form

$$f_{\chi^2}(x; n-1) = \frac{1}{2^{(n-1)/2} \Gamma((n-1)/2)} x^{(n-3)/2} e^{-x/2}, \qquad x \geq 0, \tag{3.6.11}$$

which traditionally is called the chi-square p.d.f. with $(n-1)$ degrees of freedom.

Again, it is more convenient here to tabulate the tail quantiles $q_{\chi^2}(\alpha; n)$ rather than the c.d.f.s themselves; see Table 3.6.3. Thus a C-confidence-level interval for σ_X^2 is of the form

$$\left(\frac{(n-1)S_X^2}{q_{\chi^2}((1-C)/2; n-1)}, \ \frac{(n-1)S_X^2}{q_{\chi^2}((1+C)/2; n-1)}\right) \tag{3.6.12}$$

if we decide to make symmetric cutoffs at the top and bottom of the range of the chi-square p.d.f.

Example 3.6.3 (A 99% confidence interval for $\mathrm{Var}(X)$*).* Twenty-six independent measurements of a Gaussian random quantity X resulted in the estimate $S_{26}^2 = 1.37$ for the variance $\mathrm{Var}(X)$. With $C = 0.99$, Table 3.6.3 yields

$$q_{\chi^2}((1 + 0.99)/2; 25) = q_{\chi^2}(0.995; 25) = 10.52$$

and

$$q_{\chi^2}((1 - 0.99)/2; 25) = q_{\chi^2}(0.005; 25) = 46.92.$$

Thus the 99% confidence-level interval for the variance σ_X^2 is

$$\left(\frac{25 \cdot 1.37}{46.92}, \ \frac{25 \cdot 1.37}{10.52}\right) = (0.72, \ 3.25).$$

Table 3.6.3 Tail quantiles $q_{\chi^2}(\alpha; n)$ of the chi-square distribution

$n \backslash \alpha$	0.9950	0.9900	0.9750	0.9500	0.9000	0.1000	0.0500	0.0250	0.0100	0.0050
1	0.000	0.000	0.001	0.004	0.016	2.706	3.843	5.025	6.637	7.882
2	0.010	0.020	0.051	0.103	0.211	4.605	5.992	7.378	9.210	10.597
3	0.072	0.115	0.216	0.352	0.584	6.251	7.815	9.348	11.344	12.937
4	0.207	0.297	0.484	0.711	1.064	7.779	9.488	11.143	13.277	14.860
5	0.412	0.554	0.831	1.145	1.160	9.236	11.070	12.832	15.085	16.748
6	0.676	0.872	1.237	1.635	2.204	10.645	12.592	14.440	16.812	18.548
7	0.989	1.239	1.690	2.167	2.833	12.17	14.067	16.012	18.474	20.276
8	1.344	1.646	2.180	2.733	3.490	13.362	15.507	17.534	20.090	21.954
9	1.735	2.088	2.700	3.325	4.168	14.684	16.919	19.022	21.665	23.587
10	2.156	2.558	3.247	3.940	4.865	15.987	18.307	20.483	23.209	25.188
11	2.603	3.053	3.816	4.575	5.578	17.275	19.675	21.920	24.724	26.755
12	3.074	3.571	4.404	5.226	6.304	18.549	21.026	23.337	26.217	28.300
13	3.565	4.107	5.009	5.892	7.041	19.812	22.362	24.735	27.687	29.817
14	4.075	4.660	5.629	6.571	7.790	21.064	23.685	26.119	29.141	31.319
15	4.600	5.229	6.262	7.261	8.547	22.307	24.996	27.488	30.577	32.799
16	5.142	5.812	6.908	7.962	9.312	23.542	26.296	28.845	32.000	34.267
17	5.697	6.407	7.564	8.682	10.085	24.769	27.587	30.190	33.408	35.716
18	6.265	7.015	8.231	9.390	10.865	25.989	28.869	31.526	34.805	37.156
19	6.843	7.632	8.906	10.117	11.651	27.203	30.143	32.852	36.190	38.580
20	7.434	8.260	9.591	10.851	12.443	28.412	31.410	34.170	37.566	39.997
21	8.033	8.897	10.283	11.591	13.240	29.615	32.670	35.479	38.930	41.399
22	8.643	9.542	10.982	12.338	14.042	30.813	33.924	36.781	40.289	42.796
23	9.260	10.195	11.688	13.090	14.848	32.007	35.172	38.075	41.637	44.179
24	9.886	10.856	12.401	13.848	15.659	33.196	36.415	39.364	42.980	45.558
25	10.519	11.523	13.120	14.611	16.473	34.381	37.652	40.646	44.313	46.925
26	11.160	12.198	13.844	15.379	17.292	35.563	38.885	41.923	45.642	48.290
27	11.807	12.878	14.573	16.151	18.114	36.741	40.113	43.194	46.962	49.642
28	12.461	13.565	15.308	16.928	18.939	37.916	41.337	44.461	48.278	50.993
29	13.120	14.256	16.147	17.708	19.768	39.087	42.557	45.772	49.586	52.333
30	13.787	14.954	16.791	18.493	20.599	40.256	43.773	46.979	50.892	53.672
31	14.457	15.655	17.538	19.280	21.433	41.422	44.985	48.231	52.190	55.000
32	15.134	16.362	18.291	20.072	22.271	42.585	46.194	49.480	53.486	56.328
33	15.814	17.073	19.046	20.866	23.110	43.745	47.400	50.724	54.774	57.646
34	16.501	17.789	19.806	21.664	23.952	44.903	48.602	51.966	56.061	58.964
35	17.191	18.508	20.569	22.465	24.796	46.059	49.802	53.203	57.340	60.272
36	17.887	19.233	21.336	23.269	25.643	47.212	50.998	54.437	58.619	61.581
37	18.584	19.960	22.105	24.075	26.492	48.363	52.192	55.667	59.891	62.880
38	19.289	20.691	22.878	24.884	27.343	49.513	53.384	56.896	61.162	64.181
39	19.994	21.425	23.654	25.695	28.196	50.660	54.572	58.119	62.462	65.473
40	20.706	22.164	24.433	26.509	29.050	51.805	55.758	59.342	63.691	66.766

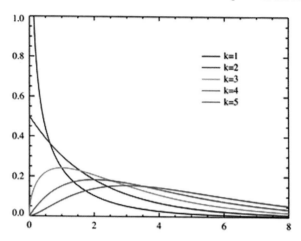

Fig. 3.6.2 Chi-square p.d.f.s with (top to bottom) 1, 2, 3, 4, and 5 degrees of freedom (from http://en.wikipedia.org/wiki/chi-square-distribution)

The interval is relatively large because the confidence level demanded is very high. Note that it is not symmetric about the estimated value $S_{26}^2 = 1.37$.

Remark 3.6.2 (Asymmetry of the chi-square distribution). Both the standard Gaussian and Student's t distribution are symmetric about the origin; their p.d.f.s are even functions. For that reason, to construct confidence intervals for them at a given (high) confidence level, it is sufficient to know their tail quantiles only for small tail probabilities. However, the chi-square distribution is asymmetric; see, Fig. 3.6.2 Thus the tables need to contain tail quantiles for both small and large (close to 1) tail probabilities. This need is on display in the above Example 3.6.3.

3.7 Problems, Exercises, and Tables

Use *Mathematica, Maple,* or *Matlab* as needed throughout this and other problem sections.

3.7.1. Plot the c.d.f.s of binomial random quantities X with $p = 0.21$ and $n = 5, 13, 25$. Calculate probabilities that X takes values between 1.3 and 3.7. Repeat the same exercise for $p = 0.5$ and $p = 0.9$.

3.7.2. Calculate the probability that a random quantity uniformly distributed over the interval $[0, 3]$ takes values between 1 and 3. Do the same calculation for the exponentially distributed random quantity with parameter $\mu = 1.5$, and the Gaussian random quantity with parameters $\mu = 1.5, \sigma^2 = 1$.

3.7.3. Prove that $\gamma \Gamma(\gamma) = \Gamma(\gamma + 1)$ and that $\Gamma(n) = (n - 1)!$ Use the integration-by-parts formula. Verify analytically that $\Gamma(1/2) = \sqrt{\pi}$. Use the idea employed in Example 3.1.6 to prove that the standard Gaussian density is normalized. Then calculate moments of order n of the standard Gaussian distribution.

3.7.4. The p.d.f. of a random variable X is expressed by the quadratic function $f_X(x) = ax(1 - x)$, for $0 < x < 1$, and is zero outside the unit interval. Find a from the normalization condition and then calculate $F_X(x)$, $\mathbf{E}X$, $\text{Var}(X)$, $\text{Std}(X)$, the nth central moment, and $\mathbf{P}(0.4 < X < 0.9)$. Graph $f_X(x)$ and $F_X(x)$.

3.7.5. Find the c.d.f and p.d.f. of the random quantity $Y = X^3$, where X is uniformly distributed on the interval $[1, 3]$.

3.7.6. Find the c.d.f and p.d.f. of the random quantity $Y = \tan X$, where X is uniformly distributed over the interval $(-\pi/2, \pi/2)$. Find a physical (geometric) interpretation of this result. Show that the second moment of Y (and thus variance) is infinite, and that the expectation $\mathbf{E}(Y)$ is not well defined despite the symmetry of the p.d.f. about zero. Also, see Problem 3.7.28.

3.7.7. Verify that $\text{Var}(X) = \mathbf{E}X^2 - (\mathbf{E}X)^2$; see formula (3.2.6).

3.7.8. Calculate the expectation and the variance of the binomial distribution from Example 3.1.2.

3.7.9. Calculate the expectation and the variance of the Poisson distribution from Example 3.1.3.

3.7.10. Calculate the expectation, variance, and nth moment of the exponential distribution from Example 3.1.5.

3.7.11. Calculate the nth central moment of the Gaussian distribution from Example 3.1.6.

3.7.12. Derive the formula for the binomial distribution from Example 3.1.2 relying on the observation that it is the distribution of the sum of n independent and identically distributed Bernoulli random quantities. Show that if $p = \mu/n$ and $n \to \infty$, then the binomial probabilities converge to the Poisson probabilities.

3.7.13. A random quantity X has an even p.d.f. $f_X(x)$ of the triangular shape shown in Fig. 3.7.1.

(a) How many parameters do you need to describe this p.d.f.? Find an explicit analytic formula for the p.d.f. $f_X(x)$ and the c.d.f. $F_X(x)$. Graph both.
(b) Find the expectation and variance of X.
(c) Let $Y = X^3$. Find the p.d.f. $f_Y(y)$ and graph it.

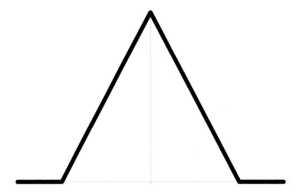

Fig. 3.7.1 A triangular p.d.f. from Problem 3.7.13

3.7.14. A discrete 2D random vector (X, Y) has the following joint p.d.f.:

$$P(X = 1, Y = 1) = \frac{2}{12}, P(X = 2, Y = 1) = \frac{1}{12}, P(X = 3, Y = 1) = \frac{1}{12},$$

$$P(X = 1, Y = 3) = \frac{2}{12}, P(X = 2, Y = 3) = \frac{4}{12}, P(X = 3, Y = 2) = \frac{2}{12}.$$

Find the marginal distributions of X and Y, their expectations and variances, as well as the covariance and correlation coefficient of X and Y. Are X and Y independent?

3.7.15. Verify the Cauchy–Schwartz inequality (3.3.18). *Hint:* Take $Z = (X - EX)/\sigma(X)$ and $W = (Y - EY/\sigma(Y)$, and consider the discriminant of the expression $E(Z + xW)^2$. The latter is quadratic in the x variable and necessarily always nonnegative, so it can have at most one root.

3.7.16. The following sample of the random vector (X, Y) was obtained: $(1, 1.7)$, $(2, 2), (5, 4.3), (7, 5.9), (9, 8), (9, 8.7)$. Produce the scatterplot of the sample and the corresponding least-squares regression line.

3.7.17. Using the table of $N(0, 1)$ c.d.f. provided at the end of this chapter, calculate $P(-1 \leq Y \leq 2)$ if $Y \sim N(0.7, 4)$.

3.7.18. Produce graphs of Student's t p.d.f. $f_T(x, n)$, for $n = 2, 5, 12, 20$, and compare them with the standard normal p.d.f.

3.7.19. Produce graphs of the chi-square p.d.f. $f_{\chi^2}(x, n)$ for $n = 2, 5, 12, 20$.

3.7.20. Find a constant $c > 0$ such that the function

$$f_X(x) = \begin{cases} c(1 + x)^{-4}, & \text{for } x > 0; \\ 0, & \text{for } x \leq 0, \end{cases}$$

is a valid p.d.f. Find $P(1/5 < X < 5)$, $E(X)$, and the p.d.f. $f_Y(y)$, of $Y = X^{1/5}$.

3.7.21. Measurements of voltage V and current I on a resistor yielded the following $n = 5$ paired data: (1.0, 2.3), (2.0, 4.1), (3.0, 6.4), (4.0, 8.5), (5.0, 10.5). Draw the scatterplot and find the regression line providing the least-squares fit for the data.

3.7.22. Independent measurements of the leakage current I on a capacitor yielded the following data: 2.71, 2.66, 2.78, 2.67, 2.71, 2.69, 2.70, 2.73 mA. Assuming that the distribution of the random quantity I is Gaussian, find the 95% confidence intervals for the expectation $\mathbf{E}I$ and the variance σ_I^2.

3.7.23. Verify that the random quantities $Z_n, n = 1, 2, \ldots$, defined in (3.5.5) have expectation 0 and variance 1.

3.7.24. Complete the following sketch of the proof of the central limit theorem from Sect. 3.5. Start with a simplifying observation (based on Problem 3.7.23) that it is sufficient to consider random quantities $X_n, n = 1, 2, \ldots$, with expectations equal to 0, and variances 1.

(a) Define $\mathcal{F}_X(u)$ as the inverse Fourier transform of the distribution of X:

$$\mathcal{F}_X(u) = \mathbf{E}e^{juX} = \int_{-\infty}^{\infty} e^{jux}\, dF_X(x).$$

Find $\mathcal{F}'_X(0)$ and $\mathcal{F}''_X(0)$. In the statistical literature $\mathcal{F}_X(u)$ is called the *characteristic function* of the random quantity X. Essentially, it completely determines the probability distribution of X via the Fourier transform (inverse of the inverse Fourier transform).

(b) Calculate $\mathcal{F}_X(u)$ for the Gaussian $N(0, 1)$ random quantity. Note the fact that its functional shape is the same as that of the $N(0, 1)$ p.d.f. This fact is the crucial reason for the validity of the CLT.

(c) Prove that, for independent random quantities X and Y,

$$\mathcal{F}_{X+Y}(u) = \mathcal{F}_X(u) \cdot \mathcal{F}_Y(u).$$

(d) Utilizing (c), calculate

$$\mathcal{F}_{\sqrt{n}(\bar{X}-\mu_X)/\mathrm{Std}(X)}(u).$$

Then find its limit as $n \to \infty$. Compare it with the characteristic of the Gaussian $N(0, 1)$ random quantity. (Hint: It is easier to work here with the logarithm of the above transform.)

3.7.25. Use the above-introduced characteristic function technique to prove that the sum of two independent Gaussian random quantities is again a Gaussian random quantity.

3.7.26. What is the probability P that a randomly selected chord is shorter than the side S of an equilateral triangle inscribed in the circle? Here are two seemingly reasonable solutions[20]

(a) A chord is determined by its two endpoints. Fix one of them to be A. For the chord to be shorter than the side S, the other endpoint must be chosen on either the arc AB or on the arc CA, and each of them is subtended by an angle of $120°$. Thus, $P = 2/3$.
(b) A chord is completely determined by its center. For the chord to be shorter than the side S, the center must lie outside the circle of radius equal to half the radius of the original circle and the same center. Hence, the probability P equals the ratio of the annular area between two circles and the area of the original circle, which is 3/4.

These two solutions are different. How is that possible?

3.7.27. Derive formulas for the c.d.f. $F_Y(y)$ and the p.d.f. $f_Y(y)$ of a transformation $Y = g(X)$ of a random quantity X, in terms of its c.d.f. $F_X(x)$ and p.d.f. $f_X(x)$ when the transforming function $y = g(x)$ is *monotonically decreasing*. Follow the line of reasoning used to derive the analogous formulas (3.1.11) and (3.1.12) for *monotonically increasing* transformations. How would you extend these formulas to transformations that are monotonically increasing on some intervals and decreasing on their complement?

3.7.28. Consider the Cauchy random quantity X defined in Remark 3.1.2. Plot its c.d.f., and then plot $X = X(\omega)$ as a function on the unit interval. Calculate the probability that X takes values between -3 and $+3$. Compare it with the similar probability for the standard Gaussian random quantity. Find and plot its p.d.f. Compare the rate of decay at $+\infty$ of the Cauchy p.d.f. with that of the $N(0, 1)$ p.d.f. Show that the expectation of the Cauchy random quantity is undefined and its variance is infinite.

3.7.29. Show that the variance estimator S_n^2 introduced in (3.6.2) is unbiased; that is, $\mathbf{E}S_n^2 = \mathrm{Var}(X)$. Also, see Problem 3.7.6.

---

[20] For more information, see Example 5.1.1. in M. Denker and W. A. Woyczyński, *Introductory Statistics and Random Phenomena: Uncertainty, Complexity and Chaotic Behavior in Engineering and Science*, Birkhäuser-Boston, Cambridge, MA, 1998.

Chapter 4
Stationary Signals

In this chapter we introduce basic concepts necessary to study the time-dependent dynamics of random phenomena. The latter will be modeled as a family of random quantities indexed by a parameter, interpreted in this book as time. The parameter may be either continuous or discrete. Depending on the context, and on the tradition followed by different authors, such families are called *random signals, stochastic processes*, or (random) *time series*. The emphasis here is on random dynamics that are *stationary*, that is, governed by underlying statistical mechanisms that do not change in time, although, of course, particular realizations of such families will be functions that vary with time. Think here about a random signal produced by the proverbial repeated coin tossing; the outcomes vary while the fundamental mechanics remain the same.

4.1 Stationarity and Autocovariance Functions

A *random* (or *stochastic*) *signal* is a time-dependent family of real-valued[1] random quantities $X(t)$. Depending on the context, one can consider random signals on the positive timeline, $t \geq 0$, on the whole timeline, $-\infty < t < \infty$, or on a finite time interval, $t_0 \leq t \leq t_1$. Also, it is useful to be able to consider random vector signals and signals with discrete time $t = \ldots, -2, -1, 0, 1, 2, \ldots$.

In this book we will restrict our attention to signals that are statistically stationary, which means that at least some of their statistical characteristics do not change in time. Several choices are possible here:

First-order strictly stationary signals. In this case the c.d.f. $F_{X(t)}(x) = \mathbf{P}(X(t) \leq x)$ does not change in time (is time-shift invariant); that is,

$$F_{X(t)}(x) = F_{X(t+\tau)}(x), \quad \text{for all} \quad t, \tau, x. \tag{4.1.1}$$

---

[1] At the end of this section we will show how the concepts discussed below should be adjusted if one considers the complex-valued stochastic signals.

W.A. Woyczyński, *A First Course in Statistics for Signal Analysis*,
DOI 10.1007/978-0-8176-8101-2_4, © Springer Science+Business Media, LLC 2011

Second-order strictly stationary signals. In this case the joint c.d.f.

$$F_{(X(t_1),X(t_2))}(x_1, x_2) = \mathbf{P}(X(t_1) \leq x_1, X(t_2) \leq x_2)$$

does not change in time; that is,

$$F_{(X(t_1),X(t_2))}(x_1, x_2) = F_{(X(t_1+\tau),X(t_2+\tau))}(x_1, x_2), \quad \text{for all} \quad t_1, t_2, \tau, x_1, x_2$$
(4.1.2)

In a similar fashion one can define the *nth-order strict stationarity* of a random signal $X(t)$ as the time-shift invariance of the *n*th-order joint c.d.f., that is, the requirement that

$$F_{(X(t_1),...,X(t_n))}(x_1, \ldots, x_n) = F_{(X(t_1+\tau),...,X(t_n+\tau))}(x_1, \ldots, x_n), \quad (4.1.3)$$

for all $t_1, \ldots, t_n, \tau, x_1, \ldots, x_n$.

Finally, a random signal $X(t)$ is said to be *strictly stationary* if, for each $n = 1, 2, \ldots$, it is *n*th-order strictly stationary.

Obviously, as n increases, verifying the *n*th-order stationarity gets more and more difficult, not to mention the practical difficulties in checking the full strict stationarity. For this reason, a more modest concept of *second-order weakly stationary signals* is useful. In this case the invariance property is demanded only of the moments of the signal up to order two. More precisely, we have the following fundamental.

Definition 4.1.1. A signal $X(t)$ is said to be second-order *weakly stationary* if its expectations and covariances are time-shift-invariant, that is, if for all t, τ,

$$\mu_X(t) \equiv \mathbf{E}[X(t)] = \mathbf{E}[X(t + \tau)] \equiv \mu_X(t + \tau), \quad (4.1.4)$$

and, for all t_1, t_2, τ, the *autocovariance function (ACvF)* is

$$\gamma_X(t_1, t_1 + \tau) \equiv \mathrm{Cov}(X(t_1), X(t_1 + \tau))$$
$$= \mathrm{Cov}(X(t_2), X(t_2 + \tau)) \equiv \gamma_X(t_1, t_2 + \tau), \quad (4.1.5)$$

where, as in Chap. 3, for a random vector (X, Y), the covariance is

$$\mathrm{Cov}(X, Y) = \mathbf{E}(X - \mu_X)(Y - \mu_Y).$$

It is a consequence of the above two conditions that, for any second-order weakly stationary signal,

$$\mu_X(t) = \mu_X = \text{const}, \quad (4.1.6)$$

and the autocovariance function depends only on the *time lag* τ and can be written as a function of a single variable:

$$\gamma_X(t, t + \tau) = \gamma_X(0, \tau) = \gamma_X(\tau), \quad (4.1.7)$$

or, equivalently,

$$\gamma_X(s,t) \equiv \gamma_X(0, t-s) = \gamma_X(t-s). \qquad (4.1.8)$$

Note that the variance of the stationary signal is also independent of time and is equal to the value of ACvF at $\tau = 0$. Indeed,

$$\text{Var}(X(t)) = \text{Cov}(X(t), X(t)) = \gamma_X(t,t) = \gamma_X(0) = \sigma_X^2 = \text{const} \qquad (4.1.9)$$

In the remainder of these lecture notes we will restrict our attention to second-order weakly stationary signals $X(t)$, which we will simply call *stationary signals*. We will analyze them assuming only the knowledge of their mean value μ_X and their autocovariance function $\gamma_X(t)$.

The following properties of the autocovariance function follow directly from its definition and the Schwartz inequality (see Sect. 3.7):

$$\gamma_X(-\tau) = \gamma_X(\tau) \qquad (4.1.10)$$

and

$$|\gamma_X(\tau)| \leq \gamma_X(0) = \sigma_X^2. \qquad (4.1.11)$$

In other words, the covariance function is even and its absolute value is dominated by its value at $\tau = 0$, where it is simply equal to the signal's variance.

Remark 4.1.1 (Autocovariance function (ACvF) vs. autocorrelation function (ACF)). You may remember that in Chap. 3 [see (3.3.19)] we defined the correlation coefficient of the random quantities X and Y as normalized covariance, that is, the covariance of X and Y divided by the product of the standard deviations of X and Y. Thus, for weakly stationary signals, the *autocorrelation function* is also dependent only on the time lag and is expressed by the formula

$$\rho_X(\tau) = \frac{\text{Cov}(X(t), X(t+\tau))}{\text{Std}(X(t))\text{Std}(X(t+\tau))} = \frac{\gamma_X(\tau)}{\gamma_X(0)}.$$

So, in view of (4.1.11), the autocorrelation function always takes values between -1 and $+1$. However, in this book we will employ only the autocovariance function, as it also contains information about the variance of the signal (as its value at $\tau = 0$), which, as we will see later on, represents the mean power of the signal. However, in the signal processing literature, one often finds the autocovariance function $\gamma_X(\tau)$, called the *autocorrelation function*, without normalizing it. So, when consulting a particular book or article, one has to make sure what definition of the ACvF is employed.

The reminder of this section is devoted to a series of examples of stationary data. The first, a real-life example shown in Fig. 4.1.1, displays a sample of a 21-channel recording of the sleep electroencephalogram (EEG), of a neonate. The duration of this multidimensional random signal is 1 min and the sampling rate is 64 Hz. This particular EEG was taken during the so-called mixed-frequency sleep stage and, in

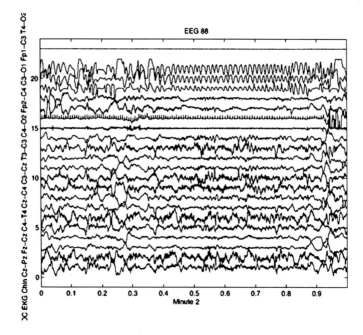

Fig. 4.1.1 A sample of a 21-channel recording of the sleep electroencephalogram (EEG) of a neonate. The duration of this multidimensional random signal is 60 s and the sampling rate is 64 Hz. (From A. Piryatinska's Ph.D. dissertation, Department of Statistics, Case Western Reserve University, Cleveland, OH, 2004)

addition to the EEG, also shows related signals such as electrocardiogram (EKG), breathing signal, eye muscle contraction signal, etc. the signal's components seem stationary for some channels while other channels seem to violate the stationarity property. This can be due to some artifacts in the recordings caused, for example, by the physical movements of the infant or by the onset of a different sleep stage [active, passive, rapid eye movement (REM), etc]. The study of EEG signals provides important information on the state of the brain's neural network and, in the case of infants, can be used to assess the maturity level of their brain. In Sect. 4.2 we will provide a method to estimate the autocovariance function for such real-life data.

Examples 4.1.1–4.1.6 provide various mathematical models of stationary signals. In those cases, the autocovariance functions can be explicitly calculated.

Example 4.1.1 (A random harmonic oscillation). Consider a signal which is a simple harmonic oscillation with nonrandom frequency $f_0 = 1/P$ but random amplitude A such that the second moment $\mathbf{E}A^2 < \infty$, and random phase Θ uniformly distributed over the period and independent of A. In other words,

$$X(t) = A\cos(2\pi f_0(t + \Theta)).$$

The signal is stationary because its mean value is

$$\mathbf{E}X(t) = \mathbf{E}A\cos 2\pi f_0(t + \Theta) = \mathbf{E}A \cdot \int_0^P \cos 2\pi f_0(t + \theta)\,\frac{d\theta}{P} = \mathbf{E}A \cdot 0 = 0,$$

and its autocovariance is

$$\gamma_X(t, t + \tau) = \mathbf{E}X(t)X(t + \tau) = \mathbf{E}[A\cos 2\pi f_0(t + \Theta) \cdot A\cos 2\pi f_0(t + \tau + \Theta)]$$

$$= \mathbf{E}A^2 \cdot \int_0^P \cos 2\pi f_0(t + \theta) \cdot \cos 2\pi f_0(t + \tau + \theta)\,\frac{d\theta}{P}$$

$$= \mathbf{E}A^2 \frac{1}{2}\left(\int_0^P \cos 2\pi f_0(t + t + \tau + 2\theta)\,\frac{d\theta}{P} + \int_0^P \cos 2\pi f_0(\tau)\,\frac{d\theta}{P}\right),$$

$$= \frac{\mathbf{E}A^2}{2}\cos 2\pi f_0(\tau),$$

where we used Table 1.3.1 as well as and the independence of the amplitude A and the phase Θ to split the expectations of the product into the product of the expectations. As a result, we see that the autocovariance $\gamma_X(t, t + \tau)$ is just a function of the time lag τ, which means the signal is stationary. Thus, the ACvF is

$$\gamma_X(\tau) = \frac{\mathbf{E}A^2}{2}\cos(2\pi f_0 \tau).$$

Example 4.1.2 (Superposition of random harmonic oscillations). In this example we consider a signal which is a sum of simple harmonic oscillations with frequencies $kf_0, k = 1, 2, \ldots, N$, random amplitudes $A_k, k = 1, 2, \ldots, N$, such that $\mathbf{E}A_k^2 < \infty$, and random phases $\Theta_k, k = 1, 2, \ldots, N$, uniformly distributed over the corresponding periods. All of the above random quantities are assumed to be independent of each other. In other words,

$$X(t) = \sum_{k=1}^{N} A_k \cos(2\pi k f_0(t + \Theta_k)).$$

In this case one can verify (see Sect. 4.3, Problems and Exercises) that the signal is again stationary and the covariance function is of the form

$$\gamma_X(\tau) = \frac{1}{2}\sum_{k=1}^{N} \mathbf{E}A_k^2 \cos(2\pi k f_0 \tau).$$

Example 4.1.3 (Discrete-time white noise). In this example the time is discrete, that is, $t = n = \ldots, -2, -1, 0, 1, 2, \ldots$, and the random signal $W(n)$ has mean zero and values at different times that are independent (uncorrelated would suffice) and identically distributed; we will denote their common variance by σ_W^2. In other words,

$$\mu_W = 0$$

and

$$\gamma_W(n, n + \tau) = \mathbf{E}(W(n)W(n + \tau)) = \begin{cases} \sigma_W^2, & \text{if } \tau = 0; \\ 0, & \text{if } \tau \neq 0. \end{cases}$$

Note that the above-defined signal is stationary because its autocovariance is indeed a function of only the time lag and can be written in the form

$$\gamma_W(n, n+\tau) = \sigma_W^2 \delta(\tau),$$

where

$$\delta(\tau) = \begin{cases} 1, & \text{if } \tau = 0; \\ 0, & \text{if } \tau \neq 0, \end{cases}$$

is the discrete-time version of the Dirac delta function which is usually called the *Kronecker delta*. This kind of signal is called *discrete-time white noise* and has mean zero and autocovariance function

$$\gamma_W(\tau) = \sigma_W^2 \delta(\tau).$$

Observe that in the definition of the white noise we did not specify the distribution of the random quantities $W(n)$. So, in principle, the white noise can have an arbitrary distribution as long as its variance is finite. In practice, the distribution in the white noise model to be employed must be determined from the detailed analysis of the physical phenomenon under consideration (or experimentation and estimation). Figure 4.1.2 shows a sample discrete-time white noise random signal $W(n), n = 1, 2, \ldots, 50$, with the W_ns all distributed uniformly on the interval $[-1/2 + 1/2]$. Hence $\mathbb{E}W_n = 0$, and $\sigma_W^2 = 1/12$.

By the *standard white noise*, we will always mean the white noise with variance $\sigma_W^2 = 1$. Thus we can standardize any white noise $W(n)$ by dividing all of its values by its standard deviation σ_W. So, in Example 4.1.1, the white noise $W(n)$ is not standard, but the white noise $W(n)/\sqrt{12}$ is.

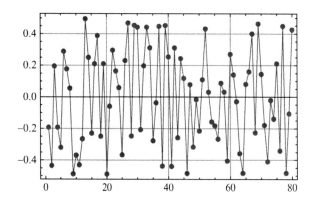

Fig. 4.1.2 A sample discrete-time white noise random signal $W(n), n = 1, 2, \ldots, 50$, with uniform distribution on the interval $[-1/2, +1/2]$, so that $\sigma_W^2 = 1/12$. For the sake of the clarity of the picture, values of $W(n)$ for consecutive integers n were joined by straight-line segments

Example 4.1.4 (Moving average of the white noise). The moving average signal $X(n)$ is obtained from the white noise $W(n)$ with variance σ_W^2 by the "windowing" procedure. The windowing procedure mixes values of the white noise, $W(n)$, $W(n-1), \ldots, W(n-q)$, in the time window of fixed width $q+1$, extending into the past, giving values with different time lags different weights, say, $b_0, b_1, \ldots, b_q$. More precisely,

$$X(n) = b_0 W(n) + b_1 W(n-1) + \cdots + b_q W(n-q).$$

You can interpret the moving average signal as a discrete-time convolution of the white noise with the windowing weight sequence. One immediately obtains that $\mu_X = 0$. Since, for independent random quantities, the variance of the sum is equal to the sum of the variances, the variance is

$$\sigma_X^2 = \sigma_W^2 \sum_{i=0}^{q} b_i^2.$$

Calculation of the autocovariance function is a little more complicated (see Sect. 4.3, Problems and Exercises) and for now we will carry it out only in the case of the window of width 2, when

$$X(n) = b_0 W(n) + b_1 W(n-1).$$

Then

$$\gamma_X(n, n+\tau) = \mathbf{E} X(n) X(n+\tau)$$

$$= \mathbf{E} \left((b_0 W(n) + b_1 W(n-1)) (b_0 W(n+\tau) + b_1 W(n+\tau-1)) \right)$$

$$= b_0^2 \mathbf{E}(W(n)W(n+\tau)) + b_0 b_1 \mathbf{E}(W(n-1)W(n+\tau))$$

$$+ b_0 b_1 \mathbf{E}(W(n)W(n+\tau-1)) + b_1^2 \mathbf{E}(W(n-1)W(n+\tau-1))$$

$$= \begin{cases} (b_0^2 + b_1^2)\sigma_W^2, & \text{if } \tau = 0; \\ b_0 b_1 \sigma_W^2, & \text{if } \tau = 1; \\ b_0 b_1 \sigma_W^2, & \text{if } \tau = -1; \\ 0, & \text{if } |\tau| > 1. \end{cases}$$

Since $\gamma_X(n, n+\tau)$ depends only on the time lag τ, the moving average signal is stationary. For the sample white noise signal from Fig. 4.1.2. the moving average signal $X(n) = 2W(n) + 5W(n-1)$ is shown in Fig. 4.1.3, and its corresponding autocovariance function,

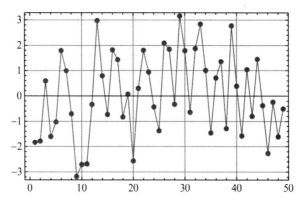

Fig. 4.1.3 Sample moving average signal $X(n) = 2W(n) + 5W(n-1)$ for the sample white noise shown in Fig. 4.1.2. Note that the moving average signal appears smoother than the original white noise. The constrained oscillations are a result of nontrivial – although short-term in this example – correlations

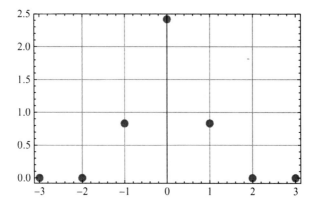

Fig. 4.1.4 Autocovariance function for the moving average signal $X(n) = 2W(n) + 5W(n-1)$. Note that the values of the signal separated by more that, one time unit are uncorrelated

$$\gamma_X(\tau) = \begin{cases} 29/12, & \text{if } \tau = 0; \\ 10/12, & \text{if } \tau = \pm 1; \\ 0, & \text{if } \tau = \pm 2, \pm 3, \ldots, \end{cases}$$

is shown in Fig. 4.1.4. Compare Figs. 4.1.3 and 4.1.4, and note that the moving average operation smoothed out the original white noise signal.

The method of determining the ACvF for a moving average signal from Example 4.1.4 can be streamlined using the fact that the ACvF of a standard white noise is the Kronecker delta. This "Kronecker delta calculus" makes it also easy to obtain the ACvF of an arbitrary infinite moving average of the white noise of the form

$$X(n) = \sum_{k=-\infty}^{\infty} b_k W(n-k), \tag{4.1.12}$$

where $W(n)$ is the standard white noise. Since $\mathbf{E}W(n)W(n+\tau) = \delta(\tau)$, which is 0 if $\tau \neq 0$ and 1 if $\tau = 0$ we have automatically that

$$
\begin{aligned}
\gamma_X(\tau) &= \mathbf{E}\left(\sum_{k=-\infty}^{\infty} b_k W(n-k) \cdot \sum_{l=-\infty}^{\infty} b_l W(n+\tau-l) \right) \\
&= \sum_{k=-\infty}^{\infty} \sum_{l=-\infty}^{\infty} b_k b_l \mathbf{E}\left(W(n-k) \cdot W(n+\tau-l) \right) \\
&= \sum_{k=-\infty}^{\infty} \sum_{l=-\infty}^{\infty} b_k b_l \delta\left((n+\tau-l) - (n-k) \right) \\
&= \sum_{k=-\infty}^{\infty} \sum_{l=-\infty}^{\infty} b_k b_l \delta\left((\tau+k) - l \right).
\end{aligned}
$$

Since $\delta(\tau - (l - k)) = 1$ if and only if $l = \tau + k$ (otherwise, it is zero), the whole double summation over the whole (k, l) lattice reduces to the single summation on the "diagonal," $l = \tau + k$, and we get the final result:

$$\gamma_X(\tau) = \sum_{k=-\infty}^{\infty} b_k b_{k+\tau}. \tag{4.1.13}$$

The variance of such a moving average signal is

$$\sigma_X^2 = \gamma_X(0) = \sum_{k=-\infty}^{\infty} b_k^2,$$

and to assure that it is finite, the sequence of coefficients, $\ldots, b_{-1}, b_0, b_1, \ldots$, must be square-summable; i.e., the condition $\sum_{k=-\infty}^{\infty} b_k^2 < \infty$ must be satisfied.

Example 4.1.5 (Random switching signal). Consider a continuous-time signal $X(t)$ switching back and forth between values $+1$ and -1 at random times. More precisely, the initial value of the signal, $X(0)$, is a random quantity with the symmetric Bernoulli distribution [i.e., $\mathbf{P}(X(0) = \pm 1) = 1/2$], and the interswitching times form a sequence $T_1, T_2, \ldots$ of independent random quantities with the identical standard exponential c.d.f.s:

$$\mathbf{P}(T_i \leq t) = 1 - e^{-t}, \qquad t > 0,$$

of mean 1. The initial random value $X(0)$ is assumed to be independent of the inter-switching times T_i. A typical sample of such a signal is shown in Fig. 4.1.5.

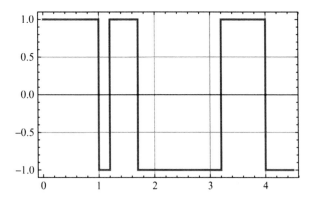

Fig. 4.1.5 A sample of the random switching signal from Example 4.1.4. The values are ± 1 and the initial value is $+1$. The interswitching times are independent and have an exponential c.d.f. of mean 1

Calculation of the mean and autocovariance function of the switching signal depends on the knowledge of the fact that such a random signal can be written in the form

$$X(t) = X(0) \cdot (-1)^{N(t)},$$

where $N(t)$ is the (nonstationary) random signal counting the number of switches up to time t; in particular, $N(0) = 0$. One can prove[2] that $N(t)$ has increments in disjoint time intervals that are statistically independent, with the distributions thereof depending only on the interval's length. More strikingly, these increments must have the Poisson probability distribution with mean equal to the interval's length, that is,

$$\mathbf{P}\left(N(t + \tau) - N(t) = k\right) = \mathbf{P}\left(T_1 + \cdots + T_k \leq \tau < T_1 + \cdots + T_{k+1}\right) = e^{-\tau} \cdot \frac{\tau^k}{k!},$$

for any $t, \tau \geq 0$, and $k = 0, 1, 2, \ldots$. Indeed,

$$\mathbf{P}(N(t) \geq k) = \mathbf{P}(T_1 + \cdots + T_k \leq t) = \int_0^t e^{-s} \frac{s^{k-1}}{(k-1)!}\, ds = e^{-t} \sum_{l=0}^k \frac{t^l}{l!},$$

because the p.d.f. of the sum of k independent standard exponential random quantities is

$$f_{T_1 + \cdots + T_k}(s) = e^{-s} \frac{s^{k-1}}{(k-1)!}, \qquad s \geq 0;$$

see Problem 4.3.8; the above integral was evaluated by repeated integration by parts.

---

[2] See, for example, O. Kallenberg, *Foundations of Modern Probability*, Springer-Verlag, New York, 1997.

Armed with this information, we can now easily complete calculations of the mean and autocovariance function of the switching signal:

$$\mu_X(t) = \mathbf{E}X(t) = \mathbf{E}X(0) \cdot \mathbf{E}(-1)^{N(t)} = 0,$$

and, for $\tau > 0$,

$$\gamma_X(t, t + \tau) = \mathbf{E}[X(t)X(t + \tau)] = \mathbf{E}X^2(0) \cdot \mathbf{E}\left[(-1)^{N(t)}(-1)^{N(t+\tau)}\right]$$
$$= 1 \cdot \mathbf{E}\left[(-1)^{2N(t)}(-1)^{N(t+\tau)-N(t)}\right] = \mathbf{E}(-1)^{N(t+\tau)-N(t)}$$
$$= \sum_{k=0}^{\infty}(-1)^k \cdot \frac{e^{-\tau}\tau^k}{k!} = e^{-2\tau}.$$

Therefore, the random switching signal $X(t)$ is stationary and, because of the symmetry property of all autocovariance functions, its ACvF is

$$\gamma_X(\tau) = e^{-2|\tau|}.$$

Remark 4.1.2 (Transition from a switching signal to the Bernoulli white noise in continuous time). Now, let us make the switching model more flexible by permitting the exponential interswitching times $T_1, T_2, \ldots$ to have mean (expected value) $\mu > 0$. That means that the common p.d.f. of the T_ks is $f_T(t) = e^{-t/\mu}/\mu$, $t \geq 0$. Recall that in Example 4.1.5 we simply assumed that $\mu = 1$. The corresponding counting, Poisson signal, $N_\mu(t)$, now has the distribution

$$\mathbf{P}(N_\mu(t) = k) = e^{-t/\mu}\frac{(t/\mu)^k}{k!}, \qquad k = 0, 1, 2, \ldots,$$

with expectation $\mathbf{E}N_\mu(t) = t/\mu$. Define the rescaled switching signal

$$X_\mu(t) = \frac{X(0)}{\sqrt{\mu}} \cdot (-1)^{N_\mu(t)},$$

with $X(0)$ independent of $N_\mu(t)$, and $\mathbf{P}(X(0) = \pm 1) = 1/2$, so that the signal $X_\mu(t)$ now switches between the values $+1/\sqrt{\mu}$ and $-1/\sqrt{\mu}$. Repeating the calculation from Example 4.1.5 in the present, general case, we obtain the following expression for its ACvF:

$$\gamma_{X_\mu}(t, t + \tau) = \mathbf{E}[X_\mu(t)X_\mu(t + \tau)] = \frac{\mathbf{E}X^2(0)}{\mu} \cdot \mathbf{E}\left[(-1)^{N_\mu(t)}(-1)^{N_\mu(t+\tau)}\right]$$
$$= \frac{1}{\mu}\mathbf{E}(-1)^{N_\mu(t+\tau)-N_\mu(t)}$$
$$= \frac{1}{\mu}\sum_{k=0}^{\infty}(-1)^k \cdot \frac{e^{-\tau/\mu}(\tau/\mu)^k}{k!} = \frac{1}{\mu}e^{-2\tau/\mu},$$

for $\tau \geq 0$. So the random switching signal $X_\mu(t)$ is stationary and its autocovariance function is

$$\gamma_{X_\mu}(\tau) = \frac{1}{\mu} e^{-2|\tau|/\mu}.$$

Now, if we let $\mathbf{E}T = \mu \to 0$, that is, if we permit the switching signal to switch more and more often, as the size of the, switches increase, then its ACvF converges to the Dirac delta impulse $\delta(t)$; see Fig. 2.1.2. So we can think about the limit of the switching signals, with $\mu \to 0$, as continuous-time white noise; it switches between $+\infty$ and $-\infty$ "infinitely often" in any finite time interval. Indeed, for any t, t_0, the expected number of switches in the time interval $[t_0, t_0 + t]$ is

$$\mathbf{E}(N_\mu(t + t_0) - N_\mu(t_0)) = \frac{t}{\mu} \to \infty, \qquad \text{as} \qquad \mu \to 0.$$

Example 4.1.6 (Solution of a stochastic difference equation). Consider a stochastic difference equation

$$X(n) = \alpha X(n-1) + \beta W(n), \qquad n = -2, -1, 0, 1, 2, \ldots,$$

where $W(n)$ is standard discrete-time white noise with $\sigma_W^2 = 1$. Observe that the above system, rewritten in the form

$$\frac{X(n) - X(n-1)}{\Delta n} = (\alpha - 1)X(n-1) + \beta W(n), \qquad n = -2, -1, 0, 1, 2, \ldots,$$

can be viewed as a discrete-time version of the stochastic differential equation

$$dX(t) = (\alpha - 1)X(t)\, dt + \beta W(t) dt,$$

where $W(t)$ represents the continuous-time version of the white noise to be discussed in later chapters and mentioned in Remark 4.1.2.

The solution of the above stochastic difference equation can be found by recursion. So

$$X(n) = \alpha(\alpha X(n-2) + \beta W(n-1)) + \beta W(n)$$
$$= \alpha^2 X(n-2) + \alpha \beta W(n-1) + \beta W(n) = \cdots$$
$$= \alpha^l X(n-l) + \sum_{k=0}^{l-1} \alpha^k \beta W(n-k),$$

for any $l = 1, 2, \ldots$. Assuming that $|\alpha| < 1$ and that $X(n-k)$ remain bounded, the first term $\alpha^k X(n-k) \to 0$ as $k \to \infty$. In that case the second term converges to the infinite sum and the solution is of the form

$$X(n) = \beta \sum_{k=0}^{\infty} \alpha^k W(n-k).$$

This is the special form of the general moving average signal appearing in (4.1.12), with the windowing sequence

$$c_k = \begin{cases} \beta\alpha^k, & \text{for } k = 0, 1, 2, \ldots; \\ 0, & \text{for } k = -1, -2, \ldots. \end{cases}$$

Hence its autocovariance function is [see (4.1.12) and (4.1.13); also, see Problem 4.3.4]

$$\gamma_X(\tau) = \sum_{k=-\infty}^{\infty} c_k c_{\tau+k} = \beta^2 \sum_{k=0}^{\infty} \alpha^k \alpha^{\tau+k} = \beta^2 \frac{\alpha^\tau}{1-\alpha^2},$$

for positive $\alpha < 1$.

Example 4.1.7 (Using moving averages to filter noise out of a signal). Consider a signal of the form

$$X(n) = \sin(0.02n) + W(n),$$

where $W(n)$ is the white noise considered in Example 4.1.3, and let $Y(n)$ be a moving average of signal $X(n)$ with the windowing sequence $b_0 = b_1 = b_2 = b_3 = b_4 = 1/5$; that is,

$$Y(n) = \frac{1}{5}X(n) + \frac{1}{5}X(n-1) + \frac{1}{5}X(n-2) + \frac{1}{5}X(n-3) + \frac{1}{5}X(n-4).$$

The values of both signals $X(n)$ and $Y(n)$, for time instants $n = 1, 2, \ldots, 750$, are shown in Fig. 4.1.6. Clearly, the moving average operation filtered some of the white noise out of the original signal, and the transformed signal appears smoother.

Remark 4.1.3 (ACvF for complex-valued signals). For complex-valued stationary signals $X(t)$, the definition of the autocovariance function has to be adjusted so that the value of the ACvF at $t = 0$ remains the variance of the signal which must be a nonnegative number. That is why taking the expectation of the simple product of values of the signal separated by the time lag τ will not do; a square of a complex number is, in general, a complex number. For that reason, for complex-valued stationary signals, the autocovariance function is defined, in the zero-mean case, by the formula

$$\gamma_X(\tau) = \mathbf{E}[X^*(t) \cdot X(t+\tau)], \qquad (4.1.14)$$

where the asterisk denotes the complex conjugate. In this case, of course, the variance is

$$\text{Var } X(t) = \mathbf{E}[X^*(t) \cdot X(t)] = \mathbf{E}|X(t)|^2 = \gamma_X(0) \geq 0.$$

Note that in the complex-valued case, the ACvF is not necessarily an even function of the time lag τ. However, we do have the equality

$$\gamma_X(-\tau) = \mathbf{E}[X^*(t) \cdot X(t-\tau)] = \big(\mathbf{E}[X^*(t-\tau) \cdot X(t)]\big)^* = \gamma_X^*(\tau). \qquad (4.1.15)$$

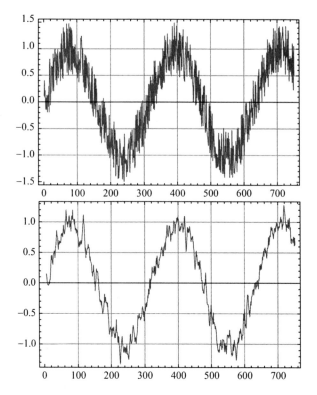

Fig. 4.1.6 (*Top*) The signal $X(n)$ from Example 4.1.6 containing a nonrandom harmonic component plus random white noise. (*Bottom*) The same signal after a smoothing, moving average operation filtered out some of the white noise. The figure shows values of both signals for times $n = 1, 2, \ldots, 750$

Example 4.1.8 (Simple complex random harmonic oscillation). Consider a complex-valued random signal represented by a simple complex exponential with a random, possibly complex-valued, amplitude A of zero mean, $\mathbf{E}A = 0$, and finite variance $\sigma_A^2 = \mathbf{E}|A|^2$:

$$X(t) = A \cdot e^{j2\pi f_0 t}.$$

Then, clearly, $\mathbf{E}\, X(t) = \mathbf{E}\, A \cdot e^{j2\pi f_0 t} = 0$, and

$$\gamma_X(\tau) = \mathbf{E}[X^*(t) \cdot X(t + \tau)] = \mathbf{E}|A|^2 \cdot e^{-j2\pi f_0 t} \cdot e^{j2\pi f_0(t+\tau)} = \sigma_A^2 \cdot e^{j2\pi f_0 \tau}.$$

This result is analogous to the result for the simple random real-valued oscillation introduced at the beginning of this section. However, in the complex-valued case, no random phase is needed to produce a stationary signal.

Example 4.1.9 (Superposition of simple complex-valued random harmonic oscillations). As in the real-valued case in Example 4.1.2, we can consider a superposition of simple complex-valued random harmonic oscillations. Let $A_1, A_2, \ldots, A_n$ be a sequence of independent (or just uncorrelated, possibly complex-valued) random amplitudes with $\mathbf{E}A_k = 0$ and finite variance $\sigma A_k^2 = \mathbf{E}|A_k|^2$. Set

$$X(t) = \sum_{k=1}^{n} A_k \cdot e^{j2\pi f_k t},$$

where $f_1, f_2, \ldots, f_n$ is a sequence of different frequencies. Then, again,

$$\mathbf{E}\, X(t) = \mathbf{E} \sum_{k=1}^{n} A_k \cdot e^{j2\pi f_k t} = \sum_{k=1}^{n} \mathbf{E}(A_k) \cdot e^{j2\pi f_k t} = 0$$

and

$$\gamma_X(\tau) = \mathbf{E}[X^*(t) \cdot X(t+\tau)] = \mathbf{E}\left(\sum_{k=1}^{n} A_k^* \cdot e^{-j2\pi f_k t} \cdot \sum_{l=1}^{n} A_l \cdot e^{j2\pi f_l (t+\tau)} \right)$$

$$= \sum_{k=1}^{n} \sum_{l=1}^{n} \mathbf{E}(A_k^* A_l) \cdot e^{-j2\pi (f_k - f_l)t} \cdot e^{j2\pi f_l \tau} = \sum_{k=1}^{n} \mathbf{E}|A_k|^2 \cdot e^{j2\pi f_k \tau},$$

because, for different k, l, the covariance $\mathbf{E}(A_k^* A_l) = \mathbf{E}(A_k^*)\mathbf{E}(A_l) = 0$.

4.2 Estimating the Mean and the Autocovariance Function; Ergodic Signals

If one can obtain multiple independent samples of the same random stationary signal, then the estimation of its parameters, the mean value and the autocovariance function can be based on procedures described in Sect. 3.6. However, very often, the only available information is a single but, perhaps, long (timewise) sample of the signal; think here about the historical temperature records at a given location, Dow Jones stock market index daily quotations over the past 10 years, or measurements of the sun spot activity over a period of time; these measurements cannot be independently repeated. Estimation of the mean and the autocovariance function of a stationary signal $X(t)$ based on its single sample is a delicate matter because the standard law of large numbers and the central limit theorem cannot be applied. So one has to proceed with caution, as we now illustrate.

Estimation of the mean μ_X. If a stationary signal $X(t)$ is sampled with the sampling interval T, that is, the known values are

$$X(0), X(T), X(2T), \ldots, X(NT), \ldots,$$

then the obvious candidate for an estimator $\hat{\mu}_X$ of the signal's mean μ_X is

$$\hat{\mu}_X(N) = \frac{1}{N} \sum_{i=0}^{N-1} X(iT).$$

This estimator is easily seen to be unbiased as

$$E[\hat{\mu}_X(N)] = \frac{1}{N} \sum_{i=0}^{N-1} E[X(iT)] = \mu_X. \qquad (4.2.1)$$

To check whether the estimator $\hat{\mu}_X(N)$ converges to μ_X as the observation interval $NT \to \infty$, that is, to check the estimator's consistency, we will take a look at the estimation error in the form of the mean-square distance (variance) between $\hat{\mu}_X(N)$ and μ_X:

$$\mathrm{Var}(\hat{\mu}_X(N)) = E[(\hat{\mu}_X - \mu_X)^2]$$

$$= \frac{1}{N^2} E\left[\sum_{i=0}^{N-1}(X(iT) - \mu_X) \sum_{k=0}^{N-1}(X(kT) - \mu_X) \right]$$

$$= \frac{1}{N^2} \sum_{i=0}^{N-1}\sum_{k=0}^{N-1} \gamma_X(iT, kT) = \frac{1}{N^2} \sum_{i=0}^{N-1}\sum_{k=0}^{N-1} \gamma_X((i-k)T)$$

$$= \frac{\sigma_X^2}{N} + \frac{2}{N} \sum_{k=0}^{N-1}\left(1 - \frac{k}{N}\right)\gamma_X(kT). \qquad (4.2.2)$$

So the error of replacing the true value μ_X by the estimator $\hat{\mu}_X$ will converge to zero, as $N \to \infty$, only if the sum in (4.2.2) increases more slowly[3] than N, i.e.,

$$\sum_{k=0}^{N-1}\left(1 - \frac{k}{N}\right)\gamma_X(kT) = o(N), \qquad \text{as} \qquad N \to \infty. \qquad (4.2.3)$$

So, for example, if the covariance function $\gamma_X(\tau)$ vanishes outside a finite interval, as was the case for finite moving averages in Example 4.1.2, then $\hat{\mu}_X$ is a consistent estimator for μ_X.

Example 4.2.1 (Consistency of the estimator $\hat{\mu}_X$ for solutions of discrete-time stochastic difference equations). Consider the solution $X(n)$ of the stochastic difference equation from Example 4.1.5. Its autocovariance function was found to be of the form

$$\gamma_X(\tau) = \beta^2 \frac{|\alpha|^\tau}{1 - \alpha^2}, \qquad |\alpha| < 1.$$

---

[3] Here we use Landau's asymptotic notation: We write that $f(x) = o(g(x))$, as $x \to x_0$, and say that $f(x)$ is little "oh" of $g(x)$ at x_0 if $\lim_{x \to x_0} f(x)/g(x) = 0$.

Since it decays exponentially as $\tau \to \infty$, the sum in (4.2.2) converges and condition (4.2.3) is satisfied. The mean-square error of replacing μ_X by the estimator $\hat{\mu}_X$ can now be controlled:

$$\text{Var}(\hat{\mu}_X(N)) = \mathbf{E}[(\hat{\mu}_X - \mu_X)^2] = \frac{\gamma_X(0)}{N} + \frac{2}{N}\sum_{k=0}^{N-1}\left(1 - \frac{k}{N}\right)\beta^2 \frac{|\alpha|^k}{1 - \alpha^2}$$

$$\leq \frac{\beta^2}{N(1 - \alpha^2)}\left(1 + 2\sum_{k=0}^{N-1}|\alpha|^k\right) = \frac{\beta^2(3 - |\alpha| - 2|\alpha|^N)}{N(1 - \alpha^2)(1 - \alpha)}.$$

Estimation of the autocovariance function $\gamma_X(\tau)$. For simplicity's sake, assume that $\mu_X = 0$, the sampling interval $T = 1$, the signal is real-valued, and that observations $X(0), \ldots, X(N)$ are given. The natural candidate for an estimator of the autocovariance function $\gamma_X(\tau) = \mathbf{E}X(0)X(\tau)$ is the time average:

$$\hat{\gamma}_X(\tau; N) = \frac{1}{N - \tau}\sum_{k=0}^{N-\tau-1} X(k)X(k + \tau). \qquad (4.2.4)$$

It is an unbiased estimator since, for each fixed time lag, τ,

$$\mathbf{E}[\hat{\gamma}_X(\tau, N)] = \frac{1}{N - \tau}\mathbf{E}\left[\sum_{k=0}^{N-\tau-1} X(k)X(k + \tau)\right]$$

$$= \frac{1}{N - \tau}\sum_{k=0}^{N-\tau-1}\gamma_X(\tau) = \gamma_X(\tau).$$

One can also prove that if $\gamma_X(\tau) \to 0$ sufficiently fast,[4] as $n \to \infty$, and if $\gamma_X(0) = \sigma_X^2 < \infty$, then the mean-square distance from $\hat{\gamma}_X(\tau; N)$ to $\gamma_X(\tau)$ decreases to 0 as $N \to \infty$. In other words, the estimator (4.2.4) is consistent.

Remark 4.2.1 (Ergodicity). If the estimator $\hat{\mu}_X$ is unbiased and consistent, that is,

$$\mathbf{E}\,\hat{\mu}_X(N) = \mu_X \quad \text{and} \quad \text{Var}(\hat{\mu}_X(N)) \to 0,$$

as $N \to \infty$, then one often says that the signal is *ergodic in the mean*. Note that, in general, this does not imply that for every sample path of the random signal the estimator converges to the estimated parameter. To guarantee that, for a general test function g, the time averages

$$\frac{g(X(1)) + g(X(2)) + \cdots + g(X(N))}{N}$$

---

[4] For a thorough exposition of these issues, see, for example, P.J. Brockwell and R.A. Davis, *Time Series: Theory and Methods*, Springer-Verlag, New York, 1991.

converge to $\mathbf{E}g(X(1))$, as $N \to \infty$, for (almost) every sample path of the random signal, stronger ergodicity and stricter stationarity assumptions are needed. A.I. Khinchin proved[5] (in the context of statistical mechanics) that a decay of the autocorrelation function to zero is a sufficient condition for ergodicity. A more detailed analysis of the ergodic behavior of stationary time series can be found in the above-quoted books by M. Denver and W.A. Woyczyński and by P.J. Brockwell and R.A. Davis.

Remark 4.2.2 (Confidence intervals). Under fairly weak assumptions one can show that the asymptotic distributions ($N \to \infty$) of the suitably rescaled estimators $\hat{\mu}_X(N), \hat{\gamma}_X(\tau; N)$ are asymptotically normal. Thus their confidence intervals can be constructed following the ideas discussed in Sect. 3.6.

Example 4.2.2 (Estimated autocorrelation functions of EEG signals). Figure 4.2.1 shows two samples of the central channel recording for a full-term neonate EEG (see Fig. 4.1.1, for a sample of the full 21-channel EEG). The duration of each of the samples was 3 min, and the signals were sampled at 64 Hz. The data in the top picture were recorded during the quiet sleep stage, and in the bottom picture – during the active sleep stage.

The estimated autocorrelation functions (ACFs) (not ACvFs!) for both signals were then calculated using formula (4.2.4) and are shown in Fig. 4.2.2. The example is taken from A. Piryatinska's Case Western Reserve Ph.D. dissertation mentioned in Sect. 4.1.

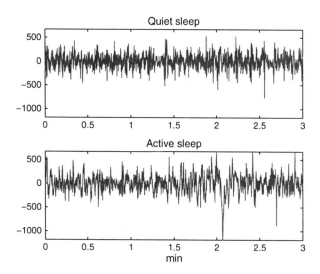

Fig. 4.2.1 (*Top*) Three-minute recording of the central channel EEG for an infant in a quiet sleep stage discussed in Example 4.2.2. (*Bottom*) Analogous recording for an active sleep stage

---

[5] See A.I. Khinchin, *Mathematical Foundation of Statistical Mechanics*, Dover Publications, New York, 1949, p. 68.

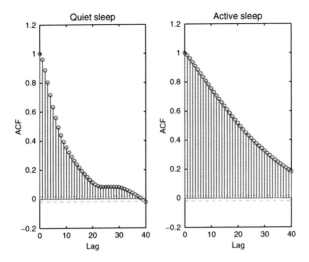

Fig. 4.2.2 (*Left*) The estimated autocovariance function (ACF) for the quiet sleep EEG signal from Fig. 4.2.1. (*Right*) Analogous estimated ACF for the active sleep stage

Note that the ACF of the active sleep signal decays much more slowly than the ACF of the quiet sleep signal, indicating the longer-range dependence structure of the former. Information on the rate of decay in EEG ACFs can then be used to automatically classify stationary segments of the EEG signals as those corresponding to different sleep stages recognized by pediatric neurologists.

4.3 Problems and Exercises

4.3.1. Consider a random signal

$$X(t) = \sum_{k=0}^{n} A_k \cos\left(2\pi k f_k(t + \Theta_k)\right),$$

where $A_0, \Theta_1, \ldots, A_n, \Theta_n$ are independent random variables of finite variance, and $\Theta_1, \ldots, \Theta_n$ are uniformly distributed on the time interval $[0, P = 1/f_0]$. Is this signal stationary? Find its mean and autocovariance functions.

4.3.2. Consider a random signal

$$X(t) = A_1 \cos 2\pi f_0(t + \Theta_0),$$

where A_1, Θ_0 are independent random variables, and Θ_0 is uniformly distributed on the time interval $[0, P/3 = 1/(3f_0)]$. Is this signal stationary? Is the signal $Y(t) = X(t) - \mathbf{E}X(t)$ stationary? Find its mean and autocovariance functions.

4.3.3. Find the mean and autocovariance functions of the discrete-time signal

$$Y(n) = 3W(n) + 2W(n-1) - W(n-2),$$

where $W(n), n = \ldots, -2, -1, 0, 1, 2, \ldots$, is the discrete-time white noise with $\sigma_W^2 = 4$; that is,

$$EW(n) = 0$$

and

$$\mathbf{E}(W(k)W(n)) = \begin{cases} 4, & \text{if } n - k = 0; \\ 0, & \text{if } n - k \neq 0. \end{cases}$$

4.3.4. Consider a general complex-valued moving average signal

$$X(n) = \sum_{k=-\infty}^{\infty} c_k W_{n-k},$$

where c_k is a complex-valued "windowing" sequence. Determine a condition on the windowing sequence that would guarantee that $X(n)$ has finite variance. $W(n)$ is the standard white noise signal with mean zero, and $\gamma_W(n) = \delta(n)$.

4.3.5. *Simulation of a discrete-time white noise with an arbitrary probability distribution.* Formula (3.1.11), $F_Y(y) = F_X(g^{-1}(y))$, describes the c.d.f. $F_Y(y)$ of the random quantity $Y = g(X)$ in terms of the c.d.f. $F_X(x)$ of the random quantity X, and a strictly increasing function $g(x)$. It also permits construction of an algorithm to produce random samples from any given probability distribution provided a random sample uniformly distributed on the interval $[0, 1]$ is given. The latter can be obtained by using the random number generator in any computing platform, see Problem 1.5.15.

Let U be random quantity U uniformly distributed on $[0,1]$ with the c.d.f.

$$F_U(u) = u, \qquad 0 \le u \le 1, \tag{4.3.1}$$

Then, for a given c.d.f. $F_Z(z)$, the random quantity $Z = F_Z^{-1}(U)$, where $F_Z^{-1}(u)$ is the function inverse to $F_Z(z)$ [that is, a solution of the equation $u = F_Z(F_Z^{-1}(u))$], has the c.d.f. $F_Z(z)$. Indeed, a simple calculation, using (4.3.1), shows that

$$\mathbf{P}(F_Z^{-1}(U) \le z) = \mathbf{P}(U \le F_Z(z)) = F_Z(z),$$

because $0 \le F_Z(z) \le 1$. So, for example, if the desired c.d.f. is exponential, with $F_Z(z) = 1 - e^{-z}, z \ge 0$, then $F_Z^{-1}(u) = -\ln(1-u), 0 \le u \le 1$, and the random quantity $Z = -\ln(1 - U)$ has the above exponential c.d.f.

The general simulation algorithm is thus as follows:

(a) Choose the sample size N, and produce a random sample, $u_1, u_2, \ldots, u_N$, uniformly distributed on $[0,1]$.
(b) Calculate the inverse function $F_Z^{-1}(u)$.
(c) Substitute the random sample, $u_1, u_2, \ldots, u_N$ into $F_Z^{-1}(u)$ to obtain the random sample

$$z_1 = F_Z^{-1}(u_1), z_2 = F_Z^{-1}(u_2), \ldots, z_N = F_Z^{-1}(u_N),$$

which has the desired c.d.f. $F_Z(z)$.

Use the above algorithm, and Problem 1.5.15, to produce and plot examples of the white noise $W(n)$ with:

(a) The standard Gaussian $N(0, 1)$ p.d.f.
(b) The double exponential p.d.f. $f_W(w) = e^{-|w|}/2$. Be careful, as its c.d.f. has a different analytic expression for positive and negative ws.
(c) The p.d.f. $f_W(w) = \sqrt{2}(\pi(1 + w^4))^{-1}$. Check that the variance is finite in this case. Start with a calculation of the corresponding c.d.f.s; a symbolic manipulation platform such as *Mathematica* is going to be great help here. Check the result graphically by plotting the histograms of the random samples against the theoretical p.d.f.s.

4.3.6. *Simulations of stationary random signals.* Using the algorithm from the Problem 4.3.5, produce simulations of stationary signals from Examples 4.1.4 and 4.1.6, using both uniformly distributed white noise, and the white noises constructed in parts (a), (b), and (c) of the above problem. Experiment with these simulations by varying parameters in the above models and changing the length of the sample of the produced discrete-time random signals.

4.3.7. Using the procedures described in Sect. 4.2, estimate the means and the auto-covariance functions (ACvF) for sample signals obtained in simulations in Problem 4.3.6. Then graphically compare the estimated and theoretical ACvFs.

4.3.8. Show that if $X_1, X_2, \ldots, X_n$ are independent, exponentially distributed random quantities with identical p.d.f.s $e^{-x}, x \geq 0$, then their sum $Y_n = X_1 + X_2 + \cdots + X_n$ has the p.d.f. $e^{-y} y^{n-1}/(n-1)!, \ y \geq 0$. Use the technique of characteristic functions (Fourier transforms) from Chap. 3. The random quantity Y_n is said to have the gamma probability distribution with parameter n. Thus the gamma distribution with parameter 1 is just the standard exponential distribution; see Example 4.1.4. Produce plots of gamma p.d.f.s with parameters $n = 2, 5, 20$, and 50. Comment on what you observe as n increases.

Chapter 5
Power Spectra of Stationary Signals

The Fourier transform $X(f)$ of the sample paths of a stationary, real-valued random signal $X(t)$ does not exist in the usual sense and analysis of the spectral contents of such signals requires a different, more subtle approach which has to rely on the concept of the *mean power* of the random signal. Only then we can investigate how it is distributed over different frequencies. The question is, of course, of fundamental importance in practical applications, as real-life signal processing devices such as measuring instruments, amplifiers, antennas, etc. transmit different frequencies with different attenuation.

5.1 Mean Power of a Stationary Signal

For stationary signals with periodic sample paths, like the superpositions of simple harmonic oscillations with random amplitudes discussed in Examples 4.1.2 and 4.1.9, the concept of the mean power is a straightforward adaptation of the power concept for periodic nonrandom signals:

$$\mathbf{E}(\mathbf{PW}_X) = \mathbf{E}\left(\frac{1}{P}\int_0^P |X(t)|^2\,dt\right) = \frac{1}{P}\int_0^P \mathbf{E}|X(t)|^2\,dt = \sigma_X^2.$$

Note that $\mathbf{PW}_X$ itself is a random quantity here. Hence, in particular, in Example 4.1.9, where

$$X(t) = \sum_{k=1}^n A_k \cdot e^{j2\pi(kf_0)t},$$

with $P = 1/f_0$, we have

$$\mathbf{E}(\mathbf{PW}_X) = \sigma_X^2 = \gamma_X(0) = \sum_{k=1}^n \mathbf{E}|A_k|^2,$$

W.A. Woyczyński, *A First Course in Statistics for Signal Analysis*, DOI 10.1007/978-0-8176-8101-2_5, © Springer Science+Business Media, LLC 2011

and the last expression provides a clear description how the mean power is distributed over different component frequencies of the signal's sample paths; the *power spectrum* in this case is discrete and the mean power carried by the frequency f_k is equal to $E|A_k|^2$.

For general stationary signals, the situation is more complicated. The mean energy $E(EN_X)$ of a stationary signal $X(t)$ over the whole timeline, that is, the expected value of energy, is infinite. Indeed, using the linearity property of expectations, we can interchange the order of taking the mean and the integration to obtain

$$E(EN_X) = E \int_{-\infty}^{\infty} X^2(t)\, dt = \int_{-\infty}^{\infty} E(X^2(t))\, dt = \int_{-\infty}^{\infty} \sigma_X^2\, dt = \infty. \quad (5.1.1)$$

However, the mean power $E(PW_X)$ of a stationary signal, taken as a limit of the mean power over finite but expanding time intervals, is always finite since

$$E(PW_X) = E \lim_{T \to \infty} \frac{1}{2T} \int_{-T}^{T} X^2(t)\, dt = \sigma_X^2 < \infty. \quad (5.1.2)$$

To find the distribution of the mean power $E(PW_X)$ over different frequencies f, we will consider a windowed signal,

$$X_T(t) = \begin{cases} X(t), & \text{for } |t| \leq T; \\ 0, & \text{otherwise}, \end{cases} \quad (5.1.3)$$

that is, the original signal restricted to the time window $-T \leq t \leq T$, of duration $2T$. Then, with the well-defined Fourier transform of the windowed signal defined by the equality

$$X_T(f) = \int_{-\infty}^{\infty} X_T(t) e^{-j2\pi ft}\, dt = \int_{-T}^{T} X(t) e^{-j2\pi ft}\, dt,$$

we can express the mean power of the original signal by the formula

$$\begin{aligned} E[PW_X] &= E\left[\lim_{T \to \infty} \frac{1}{2T} \int_{-T}^{T} X^2(t)\, dt \right] \\ &= E\left[\lim_{T \to \infty} \frac{1}{2T} \int_{-\infty}^{\infty} X_T^2(t)\, dt \right] \\ &= E\left[\lim_{T \to \infty} \frac{1}{2T} \int_{-\infty}^{\infty} |X_T(f)|^2\, df \right] \\ &= \int_{-\infty}^{\infty} \lim_{T \to \infty} \frac{E|X_T(f)|^2}{2T}\, df, \end{aligned}$$

where the Parseval equality (see Sect. 2.4) was used in the second line of the above calculation. Denoting

$$S_X(f) := \lim_{T \to \infty} \frac{\mathbf{E}|X_T(f)|^2}{2T}, \tag{5.1.4}$$

the mean power has the representation

$$\mathbf{E}(\mathbf{PW}_X) = \sigma_X^2 = \int_{-\infty}^{\infty} S_X(f) \, df. \tag{5.1.5}$$

The function $S_X(f)$ is called the *power spectral density* or, simply, the *power spectrum*, of the stationary signal $X(t)$. It shows how the mean power $\mathbf{PW}_X$ of the random stationary signal $X(t)$ is distributed over different frequencies $f, -\infty < f < \infty$. The mean power concentrated in a frequency band $f_1 < f < f_2$ is then given by the integral

$$\mathbf{PW}_X[f_1, f_2] = \int_{f_1}^{f_2} S_X(f) \, df.$$

5.2 Power Spectrum and Autocovariance Function

What makes the power spectrum $S_X(f)$ a practical tool in the analysis of random stationary signals is the fact that it is simply the Fourier transform of the signal's autocovariance function $\gamma_X(t)$. In other words,

$$S_X(f) = \int_{-\infty}^{\infty} \gamma_X(t) e^{-j2\pi f t} dt. \tag{5.2.1}$$

This fundamental property can be easily verified by direct calculation. Indeed,

$$
\begin{aligned}
S(f) &= \lim_{T \to \infty} \frac{\mathbf{E}|X_T(f)|^2}{2T} = \lim_{T \to \infty} \frac{\mathbf{E}(X_T^*(f)X_T(f))}{2T} \\
&= \lim_{T \to \infty} \frac{1}{2T} \mathbf{E} \left[\int_{-T}^{T} X^*(t) e^{2\pi j f t} dt \int_{-T}^{T} X(s) e^{-2\pi j f s} ds \right] \\
&= \lim_{T \to \infty} \frac{1}{2T} \int_{-T}^{T} \int_{-T}^{T} \mathbf{E}[X^*(t)X(s)] e^{-2\pi j f (s-t)} dt \, ds \\
&= \lim_{T \to \infty} \frac{1}{2T} \left[\int_{-T}^{T} \left(\int_{-T-s}^{T-s} \gamma_X(\tau) e^{-2\pi j f \tau} d\tau \right) ds \right] \\
&= \int_{-\infty}^{\infty} \gamma_X(\tau) e^{-2\pi j f \tau} d\tau.
\end{aligned}
$$

Given the properties of the Fourier transform, we also immediately obtain that the autocovariance $\gamma_X(\tau)$ of the signal $X(t)$ is the inverse Fourier transform of the power spectrum $S_X(f)$:

$$\gamma_X(\tau) = \int_{-\infty}^{\infty} S_X(f)e^{j2\pi f\tau}\, df. \tag{5.2.2}$$

Remark 5.2.1. What kind of functions can serve as autocovariance functions of stationary signals? Although any integrable nonnegative function,

$$S(f) \geq 0, \qquad \int_{-\infty}^{\infty} S(f)\, df < \infty,$$

can serve as a power spectrum of some stationary signal, the above formula (5.2.2) shows that for $\gamma(t)$ to be an autocovariance function of a stationary process, it must be the inverse Fourier transform of a nonnegative integrable function $S(f)$. This turns out to be a very restrictive condition. In particular, it forces $\gamma(t)$ to satisfy the following *positive-definiteness condition*:

For any positive integer N, any real number $t_1, \ldots, t_N$, and any complex numbers $z_1, \ldots, z_N$, the quadratic form is

$$\sum_{n=1}^{N}\sum_{k=1}^{N} \gamma(t_n - t_k)z_n z_k^* \geq 0.$$

Indeed, since $S(f) \geq 0$,

$$\sum_{n=1}^{N}\sum_{k=1}^{N} \gamma(t_n - t_k)z_n z_k^* = \sum_{n=1}^{N}\sum_{k=1}^{N} \int_{-\infty}^{\infty} S_X(f)e^{j2\pi f(t_n - t_k)}\, df\, z_n z_k^*$$

$$= \int_{-\infty}^{\infty} S_X(f) \sum_{n=1}^{N}\sum_{k=1}^{N} \left(z_n e^{j2\pi f t_n}\right) \cdot \left(z_k e^{j2\pi f t_k}\right)^*\, df$$

$$= \int_{-\infty}^{\infty} S_X(f) \left| \sum_{n=1}^{N} z_n e^{j2\pi f t_n} \right|^2\, df \geq 0.$$

Actually, the positive-definiteness condition is *necessary and sufficient* of a function to be an ACvF. This result is known as *Bochner's theorem*. The practical lesson is that one cannot pick examples of ACvF's off the top of one's head. There are numerous criteria guaranteeing that a given function actually is positive definite. For example, one can prove that if $\gamma(\tau)$ is even, and decreasing and convex on the positive half-line [like $\gamma(t) = e^{-|\tau|}$], then it is positive definite; see the bibliography on Fourier analysis provided at the end of this book.

Estimation of the power spectrum $S_X(f)$. For simplicity's sake, assume that the signal is real-valued and that the observations $X(0), \ldots, X(N)$ are made at discrete sampling times, $t = 0, 1, 2, \ldots, N$. To estimate the spectrum, the natural way to proceed is to replace the theoretical ACvF $\gamma_X(\tau)$ in (5.2.1) by the estimated ACvF $\hat{\gamma}_X(\tau; N)$ given by formula (4.2.4) and replacing the integral by the finite sum. This yields the estimator

$$\hat{S}_X(f; N) = \sum_{n=-(N-1)}^{N-1} \hat{\gamma}_X(|\tau|; N) e^{-j2\pi f\tau}.$$

A direct discretization of the defining formula (5.1.4) immediately gives another estimator for the power spectrum:

$$I_N(f) := \frac{1}{N} \left| \sum_{n=1}^{N} X(n) e^{-j2\pi fn} \right|^2. \tag{5.2.3}$$

For large N, $\hat{S}_X(f; N) \approx I_N(f)$ (see Sect. 9.2), and the random quantity $I_N(f)$ is usually called the *periodogram* of the sampled signal $X(t)$ based on a sample of size N.

Observe that if we have a concrete, discrete-time sample,

$$X(1) = x_1, \ldots, X(N) = x_N$$

of the signal $X(t)$, then the (nonrandom) sum inside the modulus of the periodogram formula,

$$\sum_{n=1}^{N} x_n e^{-j2\pi fn},$$

is a finite Fourier (complex-valued) trigonometric polynomial with coefficients $x_1, \ldots, x_N$. It is a periodic function of f with period $P = 1$, so the periodogram $I_N(f)$ needs to be studied only for f in the interval $[0, 1]$ (or any other interval of length 1, such as, e.g., $[-1/2 + 1/2]$). In view of the Parseval formula (2.1.12) for Fourier series,

$$\int_0^1 I_N(f)\, df = \int_0^1 \frac{1}{N} \left| \sum_{n=1}^{N} x_n e^{-j2\pi fn} \right|^2 df = \frac{1}{N} \sum_{n=1}^{N} x_n^2.$$

The expression on the right is, of course, the average power (the energy per unit time) of the sample signal $x_1, \ldots, x_N$. Thus the above formula shows that, indeed, the periodogram $I_N(f)$ gives the correct distribution of the average power of the sample signal over the frequencies $f \in [0, 1]$.

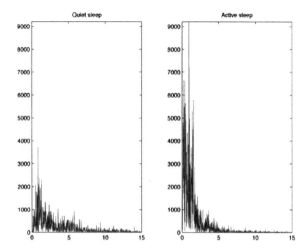

Fig. 5.2.1 *Left:* Periodogram of the neonatal quiet sleep EEG signal from Fig. 4.2.2. *Right:* Analogous periodogram for the active sleep stage. (From A. Piryatinska's 2004 Case Western Reserve University Ph.D. dissertation)

Let us return now to two samples of neonatal sleep signals displayed in Fig. 4.2.1. Their estimated autocovariance functions were shown in Fig. 4.2.2. Their periodograms have been calculated using formula (5.2.3) and are reproduced in Fig. 5.2.1. Since the signal was recorded at the sampling frequency of 64 Hz, and the duration of each recording was 3 min, the total number of sample points is $N = 192$. The reader will notice that the periodogram is quite noisy and, perhaps, should be smoothed out to better reflect the true spectrum of the random signal. Nevertheless, a comparison of these rough spectra for quiet sleep and active sleep segments clearly shows that the active sleep signal shows a bigger concentration of the spectrum at low frequencies than the quite sleep signal.

Example 5.2.1 (Simple random harmonic oscillation). In this case the random signal is of the form

$$X(t) = A\cos(2\pi f_0(t + \Theta)),$$

where the random amplitude A has zero mean, $\mathbf{E}A = 0$, and the finite variance $\mathbf{E}A^2 < \infty$. The random phase Θ is independent of A and uniformly distributed on the interval $[0, P]$ with $P = 1/f_0$. In Chap. 4 we calculated that the autocovariance function for this signal is

$$\gamma_X(\tau) = \frac{\mathbf{E}|A|^2}{2}\cos(2\pi f_0\tau).$$

Hence, the power spectrum of the simple random harmonic oscillation with fundamental frequency f_0 is

$$S_X(f) = \int_{-\infty}^{\infty} \gamma_X(\tau) e^{-2\pi j f \tau} d\tau$$

$$= \int_{-\infty}^{\infty} \frac{E|A|^2}{2} \frac{e^{j2\pi f_0 \tau} + e^{-2\pi j f_0 \tau}}{2} e^{-j2\pi f \tau} d\tau$$

$$= \frac{E|A|^2}{4} \Big(\delta(f - f_0) + \delta(f + f_0) \Big),$$

because the inverse Fourier transform of $\delta(f - f_0)$ is

$$\int_{-\infty}^{\infty} \delta(f - f_0) e^{2\pi j f \tau} df = e^{2\pi j f_0 \tau}.$$

Example 5.2.2 (Superposition of random harmonic oscillations (random periodic signal)). The signal is of the form

$$X(t) = \sum_{k=1}^{N} A_k \cos(2\pi k f_0(t + \Theta_k)),$$

where the zero-mean amplitudes $A_1, \ldots, A_N$ and phases $\Theta_1, \ldots, \Theta_N$ are all independent random quantities and $\Theta_1, \ldots, \Theta_N$ are uniformly distributed on the interval $[0, P]$, $P = 1/f_0$. The autocovariance function of this signal is

$$\gamma_X(\tau) = \sum_{k=1}^{N} \frac{E|A_k|^2}{2} \cos(2\pi k f_0 \tau),$$

and, arguing as in Example 5.2.1, the power spectrum is a linear combination of the Dirac-deltas:

$$S_X(f) = \frac{1}{4} \sum_{k=1}^{N} E|A_k|^2 \Big(\delta(f - k f_0) + \delta(f + k f_0) \Big).$$

Thus in this case the power spectrum is concentrated on the discrete frequencies $\pm f_0, \pm 2 f_0, \ldots, \pm N f_0$.

Example 5.2.3 (Band-limited noise). A stationary signal $X(t)$ is said to be a *band-limited noise* if its spectrum is

$$S_X(f) = \begin{cases} N_0, & \text{for } -f_{\max} < f < f_{\max}; \\ 0, & \text{elsewhere.} \end{cases}$$

In other words, for a band-limited noise the mean power is distributed uniformly over the frequency band $[-f_{\max}, +f_{\max}]$. The mean power of the band-limited white noise,

$$\mathbf{PW}_X = \int_{-\infty}^{\infty} S_X(f)\,df = \int_{-f_{\max}}^{f_{\max}} \mathcal{N}_0\,df = 2f_{\max}\mathcal{N}_0,$$

is finite. The autocovariance function of the band-limited white noise can be easily calculated by taking the inverse Fourier transform. Thus we obtain

$$\gamma_X(\tau) = \int_{-\infty}^{\infty} S_X(f)e^{j2\pi f\tau}\,df = \mathcal{N}_0 \int_{-f_{\max}}^{f_{\max}} e^{j2\pi f\tau}\,df$$

$$= \frac{\mathcal{N}_0}{j2\pi\tau}\left(e^{j2\pi f_{\max}\tau} - e^{-j2\pi f_{\max}\tau}\right) = \frac{\mathcal{N}_0}{\pi\tau}\sin(2\pi f_{\max}\tau).$$

Figure 5.2.2 shows both the power spectrum of a band-limited white noise and its autocovariance function, for $f_{\max} = 1$ and $\mathcal{N}_0 = 1$. Observe that, not surprisingly, as the bandwidth $2f_{\max}$ expands to infinity, the autocovariance function approaches the Dirac delta – the autocovariance function of the ideal white noise,

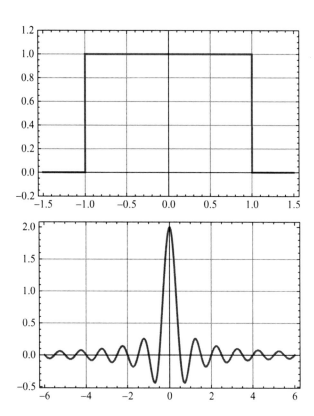

Fig. 5.2.2 *Top:* Power spectrum of the band-limited white noise $X(t)$ from Example 5.2.4. The bandwidth is $2f_{\max}$ and the mean power $\mathbf{PW}_X = 2$. *Bottom:* Autocovariance function of the above band-limited noise. Observe that as the bandwidth expands to infinity the autocovariance function approaches the Dirac delta – the autocovariance function of the ideal white noise

which will be discussed in the next example. Note that the maximum value of the autocovariance function $\gamma_X(\tau)$ is attained at $\tau = 0$ and is equal to the mean power $\mathbf{PW}_X = 2f_{max}$, which diverges to $+\infty$ as the bandwidth increases. However, $\int_{-\infty}^{\infty} \gamma_X(\tau)\,d\tau = S_X(0) = 1$, and the value of the power spectrum at zero frequency is independent of the bandwidth and remains constant.

Example 5.2.4 (The continuous-time white noise signal). By a standard white noise signal, we mean a stationary signal $W(t)$ with a totally flat power spectrum over the whole frequency range,

$$S_W(f) = 1, \qquad -\infty < f < \infty.$$

We can think about it as a limit, for $f_{max} \to \infty$, of the band-limited noise described in Example 5.2.3, but, clearly, the white noise signal is not realizable physically since its mean power is infinite:

$$\mathbf{E}(\mathbf{PW}_W) = \int_{-\infty}^{\infty} 1\,df = \infty.$$

However, it is a very useful abstraction. The Fourier transform of its autocovariance function $\gamma_W(\tau)$ must satisfy the equation

$$\int_{-\infty}^{\infty} \gamma_W(\tau)e^{-j2\pi f\tau}\,d\tau \equiv 1$$

for all $-\infty < f < \infty$, which implies that

$$\gamma_W(\tau) = \delta(\tau).$$

Loosely speaking, the above formula can be interpreted as follows: We can say that, for $t \neq s$, the white noise has "values," $X(t)$ and $X(s)$, that are uncorrelated and, for $t = s$, the covariance between $X(t)$ and $X(s)$ is infinite. This autocovariance function is thus not a true function, but its shape is not surprising if you compare it to the shape of the autocovariance function for the discrete-time white noise discussed in Chap. 4. Because of the form of its autocovariance function, the white noise is sometimes called a *delta-correlated signal*.

If a random signal $W(t)$ has the spectrum $S_W(f) \equiv \mathcal{N}_l > 0$, then we shall call $W(t)$ a white noise of amplitude $\mathcal{N}_l$.

Example 5.2.5 (Random switching signal). The random switching signal $X(t)$ discussed in Chap. 4 has the autocovariance function

$$\gamma_X(\tau) = e^{-2|\tau|}.$$

Thus its power spectral density can be directly calculated by taking the Fourier transform of the autocovariance function:

$$S_X(f) = \int_{-\infty}^{\infty} e^{-2|t|} e^{-j2\pi ft} \, dt = \int_0^{\infty} e^{-(2+j2\pi f)t} \, dt + \int_{-\infty}^0 e^{-(-2+j2\pi f)t} \, dt$$

$$= \frac{1}{2} \frac{1}{1+j\pi f} + \frac{1}{2} \frac{1}{1-j\pi f} = \frac{1}{1+(\pi f)^2}.$$

Observe that the autocovariance function decays exponentially here as the time lag increases, while the power spectrum decays only like the inverse square of the frequency when the latter goes to infinity. The situation is pictured in Fig. 5.2.3.

At this point it is worth recalling Remark 4.1.2, where we made the following observation: If instead of the above standard switching signal X, with standard, mean-one exponential interswitching times, one considers a more general switching signal X_μ, with μ (on the average) switches per unit time, then as $\mu \to 0$, its ACvF is

$$\gamma_{X_\mu}(\tau) = \frac{1}{\mu} e^{-2|\tau|/\mu} \to \delta(\tau).$$

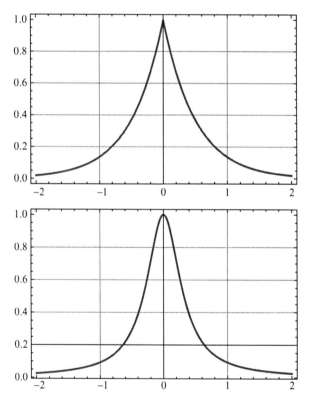

Fig. 5.2.3 *Top:* Autocovariance function of the random switching signal from Example 5.2.5. *Bottom:* The corresponding power spectrum

Thus, the standard white noise can also be seen as a limit of the general switching signals for which both the switching rate and the amplitude of the switches become larger and larger.

5.3 Power Spectra of Interpolated Digital Signals

A random signal sampled at discrete sampling time interval T_s, that is, with sampling frequency $f_s = 1/T_s$, produces a sequence of random quantities

$$\ldots, X(-2T_s), \ X(-T_s), \ X(0); X(T_s), \ X(2T_s), \ldots . \tag{5.3.1}$$

To fill in the gaps in the signal produced by discrete sampling at times nT_s, we shall interpolate the discrete signal[1] by extending its definition to other times t via the formula

$$X(t) = X(nT_s), \quad \text{for} \quad nT_s \leq t < (n+1)T_s, \tag{5.3.2}$$

and $n = \ldots, -2, -1, 0, 1, 2, \ldots$. Having extended the definition of the signal to continuous time, we can obtain its power spectrum following the method developed in Sect. 5.1. In the present case the windowed signal is of the form

$$X_N(t) = \begin{cases} X(t), & \text{for } -NT_s \leq t < NT_s; \\ 0, & \text{elsewhere}, \end{cases}$$

with the window size being $2NT_S$.

Now the mean power is

$$\mathbf{E}(\mathbf{PW}_X) = \mathbf{E} \lim_{N \to \infty} \frac{1}{2NT_s} \sum_{n=-N}^{N-1} X^2(nT_s)T_s$$

$$= \mathbf{E} \lim_{N \to \infty} \frac{1}{2NT_s} \int_{-\infty}^{\infty} |X_N(f)|^2 df \tag{5.3.3}$$

$$= \int_{-\infty}^{\infty} \lim_{N \to \infty} \frac{\mathbf{E}|X_N(f)|^2}{2NT_s} df = \int_{-\infty}^{\infty} S(f) \, df,$$

with the power spectral density

$$S(f) = \lim_{N \to \infty} \frac{\mathbf{E}|X_N(f)|^2}{2NT_s}, \tag{5.3.4}$$

and the equality in (5.3.3) resulting from Parseval's formula.

---

[1] The material in this section should be compared with analysis of the discrete and fast Fourier transforms carried out in Sect. 2.7 for nonrandom, deterministic signals.

In the next step we evaluate the Fourier transform $X_N(f)$ of the windowed interpolated signal which is needed in formula (5.3.4):

$$X_N(f) = \int_{-\infty}^{\infty} X_N(t)e^{-j2\pi ft}\, dt = \sum_{n=-N}^{N-1} \int_{nT_s}^{(n+1)T_s} X(t)e^{-j2\pi ft}\, dt$$

$$= \frac{1}{-j2\pi f} \sum_{n=-N}^{N-1} X(nT_s)\left(e^{-j2\pi f(n+1)T_s} - e^{-j2\pi fnT_s}\right)$$

$$= \frac{1 - e^{-j2\pi fT_s}}{j2\pi f} \sum_{n=-N}^{N-1} X(nT_s)e^{-j2\pi fnT_s}.$$

Substituting this result into (5.3.4), we get the following structure of the power spectrum of $X(t)$:

$$S(f) = \lim_{N\to\infty} \frac{|1 - e^{-j2\pi fT_s}|^2}{4\pi^2 f^2} \cdot \frac{\mathbf{E}\left|\sum_{n=-N}^{N-1} X(nT_s)e^{-2\pi jfnT_s}\right|^2}{2NT_s}$$

$$= \frac{1 - \cos 2\pi fT_s}{2\pi^2 f^2} \lim_{N\to\infty} \sum_{k=-N}^{N-1} \sum_{n=-N}^{N-1} \gamma_X((n-k)T_s)e^{-2\pi j(n-k)fT_s} \frac{1}{2NT_s}.$$

Changing the second summation variable by the substitution $n = m + k$, we get

$$S(f) = \frac{1 - \cos 2\pi fT_s}{2\pi^2 f^2} \lim_{N\to\infty} \sum_{k=-N}^{N-1} \sum_{m=-N-k}^{N-1-k} \gamma_X(mT_s)e^{-j2\pi mfT_s} \frac{1}{2NT_s}$$

$$= \frac{1 - \cos 2\pi fT_s}{2\pi^2 f^2 T_s^2} \cdot \sum_{m=-\infty}^{\infty} \gamma_X(mT_s)e^{-j2\pi mfT_s} T_s.$$

Hence, the power spectrum can be written as a product,

$$S(f) = S_1(f)S_2(f), \tag{5.3.5}$$

where the factor

$$S_1(f) = \frac{1 - \cos 2\pi fT_s}{2\pi^2 f^2 T_s^2} \tag{5.3.6}$$

decays to 0 at infinite frequencies ($f \to \pm\infty$) and is independent of the statistical properties of the signal [that is, of the autocovariance function $\gamma_X(nT_s)$]. The second factor,

$$S_2(f) = \sum_{m=-\infty}^{\infty} \gamma_X(mT_s)e^{-j2\pi mfT_s} T_s, \tag{5.3.7}$$

is a periodic function with period $f_s = 1/T_s$, represented by the Fourier series with coefficients given by the discrete-time autocovariance function of the discretely sampled signal.

So, if instead of the original power spectrum we consider the ratio $S(f)/S_1(f)$, then we obtain a clean relationship paralleling the symmetry of formulas for continuous-time signals:

$$\frac{S(f)}{S_1(f)} = \sum_{m=-\infty}^{\infty} \gamma_X(mT_s)e^{-j2\pi mfT_s}T_s \qquad (5.3.8)$$

and

$$\gamma_X(mT_s) = \int_{-f_s/2}^{f_s/2} \frac{S(f)}{S_1(f)}e^{j2\pi mfT_s}\,df. \qquad (5.3.9)$$

Remark 5.3.1. It is clear that all the relevant information about the spectrum of the signal sampled with the sampling interval T_s is contained in the frequency interval $(-f_s/2, +f_s/2)$. Power assigned to higher frequencies, appearing in the side "lobes" of the spectrum (see Fig. 5.3.1), is simply an artifact of the interpolation. Should we select a different interpolation scheme, the factor $S_1(f)$ responsible for the decay of the "lobes" would look different (see Sect. 5.4).

Example 5.3.1 (Interpolated moving average of the discrete-time white noise). Let the sampling interval $T_s = 1$, and let $W(n)$ be a discrete-time white noise signal $[EW(n) = 0, \; \gamma_W(\tau) = \delta(\tau)/2]$. For the moving average signal

$$Y(n) = \frac{1}{2}W(n) + \frac{1}{2}W(n-1),$$

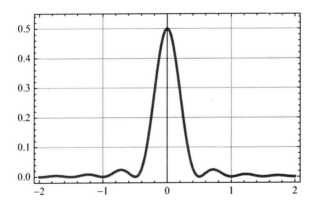

Fig. 5.3.1 Power spectrum of the interpolated moving average of the discrete-time white noise signal. The sampling rate is $f_s = 1/T_s = 1$, and the relevant spectrum is concentrated in the interval $(-f_s/2, +f_s/2)$. The side "lobes" are an artifact of the interpolation scheme

we calculated in Chap. 4 that

$$\gamma_Y(0) = 1/4, \quad \gamma_Y(\pm 1) = 1/8, \quad \gamma_Y(\tau) = 0, \quad \text{for } |\tau| \geq 2.$$

So the periodic $S_2(f)$ factor of the power spectrum of the interpolated $Y(n)$ is of the form

$$S_2(f) = \frac{1}{8} e^{j2\pi f \cdot 1} + \frac{1}{4} + \frac{1}{8} e^{-j2\pi f \cdot 1} = \frac{1}{4}(1 + \cos 2\pi f),$$

and the power spectral density itself of the interpolated Y is

$$S_Y(f) = S_1(f)S_2(f) = \frac{1 - \cos 2\pi f}{2\pi^2 f^2} \cdot \frac{1}{4}(1 + \cos 2\pi f) = \frac{1}{2}\left[\frac{\sin(2\pi f)}{2\pi f}\right]^2.$$

5.4 Problems and Exercises

5.4.1. Consider the first-order moving average signal

$$Y(n) = 4W(n) - 6W(n - 1) + 3W(n - 2),$$

where $W(n)$ is the standard discrete-time white noise signal with $\sigma_W^2 = 1$.

(a) Simulate long samples of this signal using both uniformly distributed (symmetric) and standard Gaussian white noises, and estimate its power spectrum via the periodogram formula (5.1.3). Plot it. Then smooth it out by taking its convolution with a Gaussian kernel. Plot it again.

(b) Calculate and plot the power spectrum density of Y via the "interpolation" formula in Example 5.3.1. Compare this plot with the plots obtained in part (a).

5.4.2. With $W(n)$ being the discrete-time white noise signal with $\sigma_W^2 = 5$ (either uniformly distributed and symmetric, or Gaussian), simulate long samples of the signal

$$Y(n) = W(n) + 0.5W(n - 1) - 0.3W(n - 2).$$

Derive and plot the power spectrum density of Y via both the periodogram formula (5.1.3) and the "interpolated" formula in Example 5.3.1. Follow the plan described in Exercise 5.4.1.

5.4.3. For a given window of size q, find the power spectrum density of a general moving average signal

$$Y(n) = b_0 W(n) + b_1 W(n - 1) + \cdots + b_q W(n - q),$$

where $W(n)$ is the discrete-time white noise with $\sigma_W^2 = 1$.

5.4.4. *Discrete sampling with linear interpolation.* Consider a signal X sampled at the sampling interval T_s. Its interpolation to continuous-time signal is given by the following formula:

$$X(t) = \sum_{m=-\infty}^{\infty} X(mT_s)\Lambda(t - mT_s),$$

where the interpolating kernel

$$\Lambda(t) = \begin{cases} 1 - t/T_s, & \text{for } 0 < t < T_s; \\ 1 + t/T_s, & \text{for } -T_s < t < 0; \\ 0, & \text{elsewhere.} \end{cases}$$

(a) Plot the kernel $\Lambda(t)$ and the interpolated $X(t)$ for an example of the sampled signal selected by you. Explain the interpolation effect.

(b) Demonstrate that the Fourier transform of the interpolated signal is of the form

$$X_N(f) = \sum_{m=-N}^{N} X(mT_s)e^{-2\pi jmT_s f} \Lambda(f),$$

where $\Lambda(f)$ is the Fourier transform of the kernel $\Lambda(t)$. Produce a plot of $\Lambda(f)$.

(c) Verify that the power spectrum density for the interpolated signal $X(t)$ is

$$S(f) = \lim_{N \to \infty} \frac{E|X_N(f)|^2}{(2N + 1)T_s} = \Lambda^2(f)\frac{1}{T_s} \sum_{m=-\infty}^{\infty} \gamma_X(mT_s)e^{-2\pi jmf T_s}.$$

5.4.5. A stationary signal $X(t)$ has the autocovariance function

$$\gamma_X(\tau) = 16e^{-5|\tau|}\cos 20\pi\tau + 8\cos 10\pi\tau.$$

(a) Find the variance of this signal.

(b) Find the power spectrum density of this signal.

(c) Find the value of the spectral density at zero frequency.

5.4.6. A stationary signal $X(t)$ has the spectral density of the form

$$S_X(f) = \begin{cases} 5, & \text{for } \frac{10}{2\pi} \le |f| \le \frac{20}{2\pi}; \\ 0, & \text{elsewhere.} \end{cases}$$

(a) Find the mean power of X.

(b) Find the autocovariance function of X.

(c) Find the value of the autocovariance at $\tau = 0$.

5.4.7. A stationary signal $X(t)$ has the spectral density of the form

$$S_X(f) = \frac{9}{(2\pi f)^2 + 64}.$$

At what frequency does the spectral density fall to one half of its maximal value (this value is called the *half-power bandwidth*)?

(a) Write an expression for the spectral density of a band-limited white noise Y that has the same value at zero frequency and the same mean power as X. What is its bandwidth? It is called the *equivalent noise bandwidth* of X. Compare it with the half-power bandwidth.
(b) Find the autocovariance function of the signal X.
(c) Find the autocovariance function of the signal Y.
(d) Compare the values of these two autocovariance functions at $\tau = 0$.

5.4.8. (a) Consider a solution of the stochastic differential equation described in Example 4.1.6. Take $\alpha = 0.7, \beta = 1$, and assume that the white noise $W(n)$ is Gaussian, $N(0,1)$. Produce pictures of five different trajectories of length 100 of this solution truncating the infinite series representing the solution to the first 10 terms.
(b) Use the above-generated sample signals to estimate their mean and ACvF. Plot the ACvF and compare it graphically with the theoretically derived ACvF. For better comparison, smooth out the empirical ACvFs by taking their convolution with a "nice" kernel; cf. the "moving average" technique applied in Chap. 4 to random signals themselves.
(c) Use the periodogram formula from Sect. 5.2 to estimate the power spectra of the above sample signals. Smooth them out. Compare them graphically with the theoretical power spectrum of the same signal.

5.4.9. Verify the positive-definiteness (see Remark 5.2.1) of autocovariance functions of stationary signals directly from their definition,

$$\gamma_X(\tau) = E\Big[(X(t) - E(X(t)))^* \cdot (X(t + \tau) - E(X(t)))\Big].$$

Is the stationarity condition necessary for positive-definiteness of the covariance function of $X(t)$?

Chapter 6
Transmission of Stationary Signals Through Linear Systems

Signals produced in nature are almost never experienced in their original form. Usually, we have access to them after they pass through various sensing and/or transmission devices such as a voltmeter for electric signals, an ear for acoustic signals, an eye for visual signals, a fiber-optic cable for wide-band Internet signals, and so forth. All of them impose restrictions on the signal being transmitted by attenuating different frequency components of the signal to a different degree. This process is generally called *filtering* and the devices that change the signal's spectrum are traditionally called *filters*.

A typical example here is a *band-pass filter*, which permits transmission of the components of the signal only in a certain frequency band, attenuating the frequencies in that band in a uniform fashion, but totally "killing" the frequencies outside this band. Figure 6.0.1 shows results of filtering a portion of the EEG signal from Fig. 4.1.1 through four band-pass filters with frequency bands (top to bottom) 0.5–3.5, 4–7.5, 8–12.5, and 13–17 Hz. In the neurological literature, the contents of the EEG signal within these frequency bands are traditionally called Delta, Theta, Alpha, and Beta waves, respectively.

In this chapter we study how statistical characteristics of random stationary signals are affected by transmission through linear filters. The linearity assumption means that we suppose that there is a linear relationship between the signals on the input and output of the filter. In real life it is not always the case, but the study of nonlinear filters is much more difficult than the linear theory presented below, and beyond the scope of this book.

6.1 Time-Domain Analysis

In this section we conduct the time-domain analysis of the transmission of random signals through a linear system shown schematically below:

$$X(t) \longrightarrow \boxed{h(t)} \longrightarrow Y(t).$$

W.A. Woyczyński, *A First Course in Statistics for Signal Analysis*,
DOI 10.1007/978-0-8176-8101-2_6, © Springer Science+Business Media, LLC 2011

Fig. 6.0.1 A portion of the EEG signal from Fig. 4.1.1 filtered through four band-pass filters with frequency bands (top to bottom) 0.5–3.5, 4–7.5, 8–12.5, and 13–17 Hz, respectively

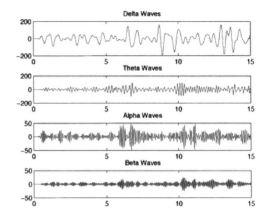

The input signal $X(t)$ is assumed to be (real-valued) random and stationary, with mean $\mu_X = EX(t)$ and autocovariance function $\gamma_X(\tau) = EX(t)X(t + \tau)$. The system is identified by a "structure" function $h(t)$, and the output signal $Y(t)$ is defined as the continuous-time moving average (convolution):

$$Y(t) = \int_{-\infty}^{\infty} X(s)h(t - s)\, ds = \int_{-\infty}^{\infty} X(t - s)h(s)\, ds. \qquad (6.1.1)$$

Note that in the case of a nonrandom Dirac delta impulse input $\delta(t)$, the nonrandom output signal is

$$y(t) = \int_{-\infty}^{\infty} \delta(s)h(t - s)\, ds = h(t - 0) = h(t).$$

For this reason the system-identifying time-domain "structure" function $h(t)$ is usually called the *impulse response function*.

The mean value of the output signal is easily calculated in terms of the input signal and of the impulse response function:

$$EY(t) = \int_{-\infty}^{\infty} E[X(t - s)]h(s)\, ds = \mu_X \int_{-\infty}^{\infty} h(s)\, ds. \qquad (6.1.2)$$

The above formula makes sense only if the last integral is well defined. For this reason we will always assume that the system is *realizable*, that is,

$$\int_{-\infty}^{\infty} |h(s)|\, ds < \infty. \qquad (6.1.3)$$

In view of (6.1.2), for realizable systems, if the input signal has zero mean, then the output signal also has zero mean:

$$\mu_X = 0 \qquad \Longrightarrow \qquad \mu_Y = 0.$$

In this situation, from now on, we will restrict our attention only to zero-mean signals.

The calculation of the autocovariance function of the output signal $Y(t)$ is a little bit more involved. Replacing the product of the integrals by the double integral, we obtain

$$\gamma_Y(\tau) = E(Y(t)Y(t+\tau))$$
$$= E\left[\int_{-\infty}^{\infty} X(t-s)h(s)\,ds \int_{-\infty}^{\infty} X(t+\tau-u)h(u)\,du\right]$$
$$= \int_{-\infty}^{\infty}\int_{-\infty}^{\infty} E\left[X(t-s)X(t+\tau-u)\right]h(s)h(u)\,ds\,du.$$

Then, in view of the stationarity assumption,

$$E\left[X(t-s)X(t+\tau-u)\right] = E\left[X(-s)X(\tau-u)\right] = \gamma_X(\tau-u+s),$$

so that, finally,

$$\gamma_Y(\tau) = \int_{-\infty}^{\infty}\int_{-\infty}^{\infty} \gamma_X(\tau-u+s)h(s)h(u)\,ds\,du. \qquad (6.1.4)$$

A system is said to be *causal* if the current values of the output depend only on the past and present values of the input. This property can be equivalently stated as the requirement that the impulse response function,

$$h(t) = 0, \quad \text{for} \quad t \le 0. \qquad (6.1.5)$$

In other words, the moving average is performed only over the past. This condition, in particular, implies that the second output integral in (6.1.1) is restricted to the positive half-line,

$$Y(t) = \int_0^{\infty} X(t-s)h(s)\,ds, \qquad (6.1.6)$$

and formula (6.1.4) for the autocovariance function takes the form

$$\gamma_Y(\tau) = \int_0^{\infty}\int_0^{\infty} \gamma_X(\tau-u+s)h(s)h(u)\,ds\,du. \qquad (6.1.7)$$

In what follows in this chapter we will consider only causal filters.

Example 6.1.1 (An integrating circuit). A standard integrating circuit with a single capacitor is shown in Fig. 6.1.1.

The impulse response function for this system is the unit step function $u(t)$ multiplied by $1/C$, where the constant C represents the capacitance of the capacitor:

$$h(s) = \frac{1}{C}u(s) = \begin{cases} 0, & \text{for } s < 0; \\ 1/C, & \text{for } s \ge 0. \end{cases}$$

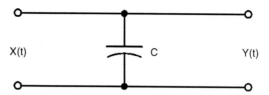

Fig. 6.1.1 A standard integrating circuit. The voltage $Y(t)$ on the output is the integral of the current $X(t)$ on the input

The output is

$$Y(t) = \frac{1}{C} \int_{-\infty}^{\infty} X(s)u(t-s)ds = \frac{1}{C} \int_{-\infty}^{t} X(s)\,ds.$$

Obviously, this system, although causal, is not realizable over the whole timeline since

$$\int_{-\infty}^{\infty} |h(t)|\,dt = \int_{0}^{\infty} \frac{1}{C}\,dt = \infty.$$

To avoid this difficulty, we need to restrict the integrating circuit to a finite time interval and assume that the adjusted impulse response function is of the form

$$h(s) = \begin{cases} 0, & \text{for } s < 0; \\ 1/C, & \text{for } 0 \le s \le T; \\ 0, & \text{for } s > T. \end{cases} \qquad (6.1.8)$$

In this situation the system is realizable and the output is

$$Y(t) = \int_{-\infty}^{\infty} X(s)h(t-s)ds = \frac{1}{C} \int_{t-T}^{t} X(s)ds.$$

The autocovariance function is equal to

$$\gamma_Y(\tau) = \int_{0}^{T} \int_{0}^{T} \gamma_X(\tau - u + s)\, h(s)\, h(u)\, ds\, du$$

$$= \frac{1}{C^2} \int_{0}^{T} \int_{0}^{T} \gamma_X(u - (\tau + s))\, ds\, du, \qquad (6.1.9)$$

because, for real-valued signals, the autocovariance function is even, $\gamma_X(-\tau) = \gamma_X(\tau)$.

Therefore, if the input signal is the standard white noise $X(t) = W(t)$ with the autocovariance function $\gamma_W(t) = \delta(t)$, and $C = 1$, then for $\tau \ge 0$, the output autocovariance function is

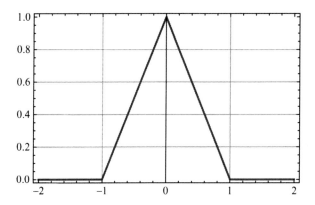

Fig. 6.1.2 The output autocovariance function $\gamma_Y(\tau)$ (6.1.10) of the integrating system (6.1.8) with $T = 1$, in the case of the standard white noise input $X(t) = W(t)$

$$\gamma_Y(\tau) = \int_0^T \int_0^T \delta(u - (\tau + s)) \, du \, ds = \int_0^T \zeta(s) \, ds,$$

where

$$\zeta(s) = \begin{cases} 0, & \text{for } \tau + s < 0; \\ 1/2, & \text{for } \tau + s = 0; \\ 1, & \text{for } 0 < \tau + s < T; \\ 1/2, & \text{for } \tau + s = T; \\ 0, & \text{for } \tau + s > T. \end{cases}$$

Hence, the autocovariance function, pictured in Fig. 6.1.2, is

$$\gamma_Y(s) = \begin{cases} 0, & \text{for } \tau < -T; \\ T - |\tau|, & \text{for } -T \le \tau \le T; \\ 0, & \text{for } \tau > T. \end{cases} \tag{6.1.10}$$

If the input signal $X(t)$ is a simple random harmonic oscillation with the autocovariance function $\gamma_X(\tau) = \cos \tau$ and, again, $C = 1$, then the output autocovariance function, several examples thereof are shown in Fig. 6.1.3, is

$$\gamma_Y(\tau) = \int_0^T \int_0^T \cos(\tau - u + s) \, ds \, du = 2 \cos \tau (1 - \cos T). \tag{6.1.11}$$

As simple as formula (6.1.9) for the output autocovariance function seems to be, the analytic evaluation of the double convolution may get tedious very quickly. Consider, for example, an input signal $X(t)$ with the autocovariance function

Fig. 6.1.3 The output
autocovariance functions
$\gamma_Y(\tau)$ (6.1.11) of the
integrating system (6.1.8)
with $T = 1, 2$, and 3 (*top* to
bottom), in the case of simple
random harmonic oscillation
input with $\gamma_X(\tau) = \cos\tau$.
Note the increasing amplitude
of $\gamma_Y(\tau)$ as T increases

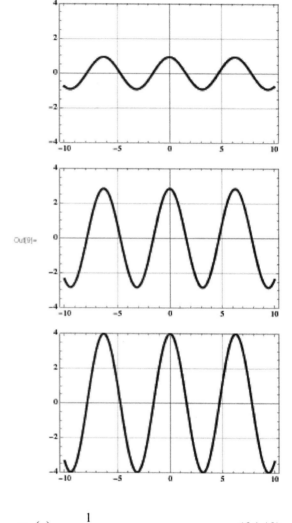

$$\gamma_X(\tau) = \frac{1}{1+\tau^2}, \qquad (6.1.12)$$

which corresponds to the exponentially decaying power spectrum (see Sect. 6.4).
In this case,

$$\gamma_Y(\tau) = \int_0^T \int_0^T \frac{1}{1+(\tau-u+s)^2}\, ds\, du$$

$$= \frac{1}{2}\Bigl(2(T-\tau)\arctan(T-\tau) - 2\tau\arctan\tau - \log(1+(T-\tau)^2)$$

$$+ \log(1+\tau^2)\Bigr)$$

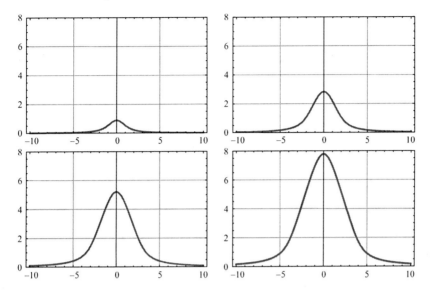

Fig. 6.1.4 The output autocovariance functions $\gamma_Y(\tau)$ (6.1.13) of the integrating system (6.1.8) with $T = 1, 2, 3,$ and 4, in the case of input with $\gamma_X(\tau) = 1/(1+\tau^2)$. Note the growing maximum and spread of $\gamma_Y(\tau)$ as T increases

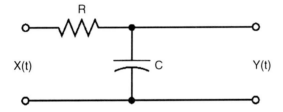

Fig. 6.1.5 A standard RC filter with the impulse response function $h(t) = (1/RC)\exp(-t/RC)\cdot u(t)$

$$+ \frac{1}{2}\bigg(-2\tau \arctan(\tau) + 2(\tau + T)\arctan(\tau + T)$$

$$+ \log(1 + \tau^2) + \log(1 + T^2 + 2T\tau + \tau^2)\bigg). \qquad (6.1.13)$$

So, even for a relatively simple autocovariance function of the input, the output autocovariance may be quite complex. And, yes, you guessed right, to avoid the tedium of paper-and-pencil calculations, we obtained the above formula using *Mathematica*. Figure 6.1.4 traces the dependence of $\gamma_Y(\tau)$ on T graphically.

Example 6.1.2. An RC filter. A standard RC filter is shown in Fig. 6.1.5.

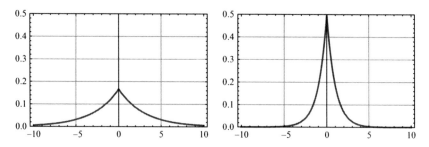

Fig. 6.1.6 The output autocovariance function $\gamma_Y(\tau)$ for the RC filter (6.1.14) with a standard white noise input with $\gamma_X(\tau) = \delta(\tau)$. The figure on the *left* shows the case of a small time constant, $RC = 1$, and the one on the *right*, the case of a larger time constant, $RC = 3$. Note the difference in the maximum and the spread of $\gamma_Y(\tau)$ in these two cases

The impulse response function of this circuit is of the form

$$h(t) = \frac{1}{RC} \exp\left(-\frac{t}{RC}\right) \cdot u(t), \tag{6.1.14}$$

where $u(t)$ is the usual unit step function, R is the electrical resistance, and C is the capacitance. The product RC represent the so-called time constant of the circuit.

In the case of the white noise input signal with $\gamma_X(\tau) = \delta(\tau)$, the output autocovariance function, for $\tau > 0$, is

$$\gamma_Y(\tau) = \int_0^\infty \int_0^\infty \delta(u - (s + \tau))h(u)h(s)\,du\,ds = \int_0^\infty h(s + \tau)h(s)\,ds$$
$$= \int_0^\infty \frac{1}{RC} e^{\frac{s+\tau}{RC}} \cdot \frac{1}{RC} e^{\frac{s}{RC}}\,ds = \frac{1}{2RC} e^{-\frac{\tau}{RC}}.$$

So

$$\gamma_Y(\tau) = \frac{1}{2RC} \exp\left(-\frac{|\tau|}{RC}\right). \tag{6.1.15}$$

The shape of the output autocovariance function for small and large values of the RC constant is shown in Fig. 6.1.6.

Remark 6.1.1 (Ornstein–Uhlenbeck stationary signals). You may have noticed that the ACvF appearing in (6.1.15) has the same exponential shape as that of the switching signal considered in Sect. 4.1, and also, in discrete time, that of the solution of a stochastic difference equation considered in the same section. However, if the input white noise in the above example has a Gaussian distribution, then the output is also Gaussian (obviously, not a switching signal, which takes only two values). A Gaussian stationary signal with the exponential ACvF (6.1.15) is traditionally called the Ornstein–Uhlenbeck signal (process), and it appears as a model in numerous physical and engineering problems, see Chap. 8 for a detailed discussion of Gaussian stationary signals.

For the simple random harmonic oscillation with autocovariance function $\gamma_X(\tau) = \cos\tau$ as the input, the output autocovariance function is

$$\gamma_Y(\tau) = \int_0^\infty \int_0^\infty \cos(\tau - u + s)\frac{1}{RC}\exp\left(\frac{-s}{RC}\right)\frac{1}{RC}\exp\left(\frac{-u}{RC}\right) ds\, du$$

$$= \frac{\cos\tau}{1 + (RC)^2}.$$

But a slightly more complex input autocovariance function,

$$\gamma_X(\tau) = e^{-2|\tau|},$$

corresponding to the switching input signal produces the output autocovariance function of the form

$$\gamma_Y(\tau) = \frac{1}{(RC)^2}\int_0^\infty \int_0^\infty e^{-|\tau-u+s|}e^{-(s+u)/(RC)}\, ds\, du$$

$$= \frac{1}{(RC)^2}\left[\int_0^\tau \int_0^\infty e^{-(\tau-u+s)}e^{-(s+u)/(RC)}\, ds\, du \right.$$

$$+ \int_\tau^\infty \left(\int_0^{u-\tau} e^{\tau-u+s}e^{-(s+u)/(RC)}\, ds\right.$$

$$\left.\left.+ \int_{u-\tau}^\infty e^{-(\tau-u+s)}e^{-(s+u)/(RC)}\, ds\right) du\right], \quad (6.1.16)$$

which, although doable (see Sect. 6.4, Problems and Exercises), is not fun to evaluate.

6.2 Frequency-Domain Analysis and System's Bandwidth

Examples provided in the preceding section demonstrated analytic difficulties related to the time-domain analysis of random stationary signals transmitted through linear systems. In many cases the analysis becomes much simpler if it is carried out in the frequency domain. For this purpose let us consider the Fourier transform $H(f)$ of the system's impulse response function $h(t)$:

$$H(f) = \int_{-\infty}^\infty h(t)e^{-2\pi jft}\, dt, \quad (6.2.1)$$

which traditionally is called the system's *transfer function*.

Now the task is to calculate the power spectrum,

$$S_Y(f) = \int_{-\infty}^\infty \gamma_Y(\tau)e^{-2\pi jf\tau}\, d\tau, \quad (6.2.2)$$

of the output signal given the power spectrum,

$$S_X(f) = \int_{-\infty}^{\infty} \gamma_X(\tau) e^{-2\pi j f \tau} \, d\tau,$$

of the input signal. Since the output autocovariance function $\gamma_Y(t)$ has been calculated in Sect. 6.1, substituting the expression obtained in (6.1.4) into (6.2.1), we get

$$S_Y(f) = \int_{-\infty}^{\infty} \left(\int_{-\infty}^{\infty} \int_{-\infty}^{\infty} \gamma_X(\tau - s + u) h(s) h(u) \, ds \, du \right) e^{-2\pi j f \tau} \, d\tau$$

$$= \int_{-\infty}^{\infty} \int_{-\infty}^{\infty} \left(\int_{-\infty}^{\infty} \gamma_X(\tau - s + u) e^{-2\pi j f (\tau - s + u)} \, d\tau \right) h(s) e^{-2\pi j f s} \, ds$$

$$\cdot h(u) e^{2\pi j f u} \, du.$$

Making the substitution $\tau - s + u = w$ in the inner integral, we arrive at the final formula:

$$S_Y(f) = S_X(f) \cdot H(f) \cdot H^*(f) = S_X(f) \cdot |H(f)|^2. \qquad (6.2.3)$$

So the output power spectrum is obtained simply by multiplying the input power spectrum by a fixed factor $|H(f)|^2$, which is called the system's *power transfer function*.

The appearance of the power transfer function, $|H(f)|^2$, in formula (6.2.3) suggests we introduce the concept of the system's bandwidth. As in the case of signals (see Sect. 5.4), several choices are possible.

The *equivalent-noise bandwidth* BW_n is defined as the cutoff frequency f_{max} of the limited-band white noise with the amplitude equal to the value of the system's power transfer function at 0 and the mean power equal to the integral of the system's power transfer function; that is,

$$2 B W_n |H(0)|^2 = \int_{-\infty}^{\infty} |H(f)|^2 \, df,$$

which gives

$$B W_n = \frac{1}{2|H(0)|^2} \int_{-\infty}^{\infty} |H(f)|^2 \, df. \qquad (6.2.4)$$

The *half-power bandwidth* $BW_{1/2}$ is defined as the frequency where the system's power transfer function declines to one half of its maximum value, which is always equal to $|H(0)|^2$. Thus it is obtained by solving, for an unknown $BW_{1/2}$, the equation

$$|H(BW_{1/2})|^2 = \frac{1}{2} |H(0)|^2. \qquad (6.2.5)$$

Obviously, the above bandwidth concepts make the best sense for lowpass filters, that is, in the case when the system's power transfer function has a distinctive maximum at 0, dominating its values elsewhere. But for other systems, such as bandpass filters, similar bandwidth definitions can be easily devised.

Example 6.2.1. An RC filter. Recall that in this case the impulse response function is given by

$$h(t) = \frac{1}{RC}e^{-\frac{t}{RC}} \cdot u(t).$$

So the transfer function is

$$H(f) = \int_{-\infty}^{\infty} h(t)e^{-2\pi jft}\, dt = \int_{0}^{\infty} \frac{1}{RC}e^{-\frac{t}{RC}}e^{-2\pi jft}\, dt = \frac{1}{1 + 2\pi jRCf}$$

and, consequently, the power transfer function is

$$|H(f)|^2 = \frac{1}{1 + 2\pi jRCf} \cdot \frac{1}{1 - 2\pi jRCf} = \frac{1}{1 + (2\pi RCf)^2}. \qquad (6.2.6)$$

Several examples of this function are shown in Fig. 6.2.1. The half-power bandwidth of the RC filter is easily computable from the equation

$$\frac{1}{1 + (2\pi RC(BW_{1/2}))^2} = \frac{1}{2},$$

which gives

$$BW_{1/2} = \frac{1}{2\pi RC}.$$

The bandwidth decreases hyperbolically with the increase of the RC constant.

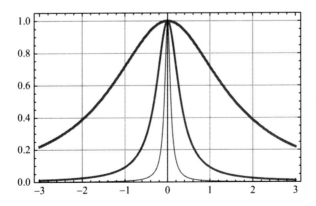

Fig. 6.2.1 Power transfer functions $|H(f)|^2 = 1/(1 + (2\pi RCf)^2)$ for the RC filter discussed in Example 6.2.1, with the RC constants 0.1 (*thick line*), 0.5 (*medium line*), and 2.0 (*thin line*). The half-power bandwidths $BW_{1/2}$ are, respectively, 1.6, 0.32, and 0.08

The output power spectra for an RC filter are thus easily evaluated. In the case of the standard white noise input with $S_X(f) \equiv 1$, the output power spectrum is

$$S_Y(f) = \frac{1}{1 + (2\pi RC f)^2}.$$

If the input signal is a random oscillation with the power spectrum

$$S_X(f) = \frac{A_0^2}{2}\Big(\delta(f - f_0) + \delta(f + f_0)\Big),$$

then the output power spectrum is

$$S_Y(f) = \frac{A_0^2}{2}\Big(\delta(f - f_0) + \delta(f + f_0)\Big) \cdot \frac{1}{1 + (2\pi RC f)^2}.$$

If the input is a switching signal with the power spectrum

$$S_X(f) = \frac{1}{1 + (af)^2},$$

then the output power spectrum is

$$S_Y(f) = \frac{1}{1 + (af)^2} \cdot \frac{1}{1 + (2\pi RC f)^2}.$$

Example 6.2.2. Bandwidth of the finite-time integrating circuit. Let us calculate the bandwidths BW_n and $BW_{1/2}$ for the finite-time integrator with the impulse response function

$$h(t) = \begin{cases} 1, & \text{for } 0 \le t \le T; \\ 0, & \text{elsewhere.} \end{cases}$$

In this case the transfer function is

$$H(f) = \int_0^T e^{-2\pi jf t}\, dt = \frac{1}{2\pi jf}\Big(1 - e^{-2\pi jf T}\Big),$$

so that the power transfer function is

$$|H(f)|^2 = \frac{(1 - e^{-2\pi jf T})(1 - e^{2\pi jf T})}{(2\pi f)^2} = \frac{2(1 - \cos 2\pi f T)}{(2\pi f)^2}. \tag{6.2.7}$$

Finding the integral of the power transfer function directly is a little tedious, but, fortunately, by Parseval's formula,

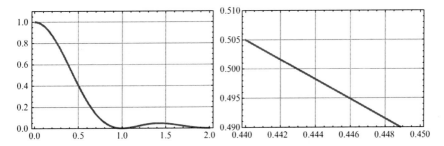

Fig. 6.2.2 *Top:* Power transfer function (6.2.7) of the finite-time integrating circuit with $T = 1$. *Bottom:* Magnified portion of the power transfer function for f between 0.44 and 0.45. This graphical analysis gives the half-power bandwidth $BW_{1/2} = 0.443$

$$\int_{-\infty}^{\infty} |H(f)|^2 \, df = \int_{-\infty}^{\infty} h^2(t) \, dt = \int_0^T dt = T$$

and

$$H(0) = \int_0^T h(t) \, dt = T.$$

Thus the equivalent-noise bandwidth (6.2.4) is

$$BW_n = \frac{1}{2T^2} \cdot T = \frac{1}{2T}.$$

Finding the half-power bandwidth requires solving equation (6.2.5),

$$\frac{2(1 - \cos 2\pi (BW_{1/2})T)}{(2\pi (BW_{1/2}))^2} = \frac{T^2}{2},$$

which can be done only numerically. Indeed, a quick graphical analysis (see Fig. 6.2.2), for $T = 1$, gives the half-power bandwidth $BW_{1/2} = 0.443$, slightly less than the corresponding equivalent-noise bandwidth $BW_{eqn} = 0.500$.

6.3 Digital Signal, Discrete-Time Sampling

In this section we will take a look at the transmission of random stationary signals through linear systems when the signals are sampled at discrete times with the sampling interval T_s. The system can be schematically represented as follows:

$$X(nT_s) \longrightarrow \boxed{h(nT_s)} \longrightarrow Y(nT_s).$$

The input signal now forms a stationary random sequence,

$$X(nT_s), \quad n = \ldots - 1, 0, 1, \ldots, \tag{6.3.1}$$

and the output signal,

$$Y(nT_s), \quad n = \ldots - 1, 0, 1, \ldots, \tag{6.3.2}$$

is produced by the discrete-time convolution of the input signal $X(nT_s)$ with the discrete-time impulse response sequence $h(nTs)$:

$$Y(nT_s) = \sum_{i=-\infty}^{\infty} X(iT_s)h(nT_s - iT_s)T_s. \tag{6.3.3}$$

In the discrete-time case, the realizability condition is

$$\sum_{n=-\infty}^{\infty} |h(nT_s)| < \infty,$$

and the causality condition means that

$$h(nT_s) = 0, \quad \text{for } n < 0.$$

With discrete-time inputs and outputs, the autocovariance functions are just discrete sequences and are defined by the formulas

$$\gamma_X(kT_s) = \mathbf{E}(X(nT_s)X(nT_s + kT_s)), \quad \gamma_Y(kT_s) = \mathbf{E}(Y(nT_s)Y(nT_s + kT_s)).$$

Then a direct application of (6.3.3) yields the following formula for the output autocovariance sequence as a function of the input autocovariance sequence and the impulse response sequence:

$$\gamma_Y(kT_s) = \sum_{l=-\infty}^{\infty} \sum_{i=-\infty}^{\infty} \gamma_X(kT_s - lT_s + iT_s)h(lT_s)h(iT_s)T_s^2. \tag{6.3.4}$$

To move into the frequency domain, one can either directly apply the discrete or fast Fourier transform or, as in Sect. 6.3, use the straight continuous-time Fourier transform technique, assuming that both the signal and the impulse response function have been interpolated by constants between sampling points. We will follow the latter approach. So, using formula (5.3.5), we get

$$S_X(f) = S_1(f) \cdot S_{2,X}(f), \tag{6.3.5}$$

with

$$S_{2,X}(f) = \sum_{m=-\infty}^{\infty} \gamma_X(mT_s)e^{-j2\pi m f T_s} T_s$$

and

$$S_Y(f) = S_1(f) \cdot S_{2,Y}(f), \tag{6.3.6}$$

with

$$S_{2,Y}(f) = \sum_{m=-\infty}^{\infty} \gamma_Y(mT_s)e^{-j2\pi mfT_s}T_s$$

and

$$S_1(f) = \frac{1 - \cos 2\pi f T_s}{2\pi^2 f^2 T_s^2}.$$

Remember that all the relevant information about the discrete sampled signal is contained in the frequency interval $(-f_s/2, f_s/2)$ (see Remark 5.3.1). The transfer function of this system is

$$H(f) = \int_{-\infty}^{\infty} h(t)e^{-j2\pi ft}\, dt = \sum_{k=-\infty}^{\infty} h(kT_s)\int_{kT_s}^{(k+1)T_s} e^{-j2\pi ft}\, dt$$

$$= \frac{1 - e^{j2\pi fT_s}}{-j2\pi f T_s} \sum_{k=-\infty}^{\infty} h(kT_s)e^{-j2\pi fkT_s}T_s, \tag{6.3.7}$$

so that the power transfer function is

$$|H(f)|^2 = \frac{1 - \cos 2\pi f T_s}{2\pi^2 f^2 T_s^2} \sum_{k=-\infty}^{\infty}\sum_{n=-\infty}^{\infty} h(kT_s)h(nT_s)e^{-j2\pi f(k-n)T_s}T_s^2. \tag{6.3.8}$$

Again, all the relevant information about the discrete power transfer function is contained in the frequency interval $(-f_s/2, f_s/2)$ (see Remark 5.3.1).

Finally, since we already know from Sect. 6.2 that

$$SY(f) = |H(f)|^2 S_X(f),$$

we also get from (6.3.5) and (6.3.6) that

$$S_{2,Y}(f) = |H(f)|^2 S_{2,X}(f) \tag{6.3.9}$$

or, equivalently,

$$\sum_{m=-\infty}^{\infty} \gamma_Y(mT_s)e^{-j2\pi mfT_s}T_s = |H(f)|^2 \cdot \sum_{m=-\infty}^{\infty} \gamma_X(mT_s)e^{-j2\pi mfT_s}T_s.$$

$$\tag{6.3.10}$$

Example 6.3.1 (Autoregressive moving average signal (ARMA)). We now take the sampling period $T_s = 1$ and the output $Y(n)$ determined from the input $X(n)$ via the autoregressive moving average scheme with parameters p and q [in short, $ARMA(p,q)$]:

$$Y(n) = \sum_{l=0}^{q} b(l)X(n-l) - \sum_{l=1}^{p} a(l)Y(n-l). \tag{6.3.11}$$

Defining $a(0) = 1$, we can then write

$$\sum_{l=0}^{p} a(l)Y(n-l) = \sum_{l=0}^{q} b(l)X(n-l).$$

Since the Fourier transform of the convolution is a product of Fourier transforms, we have

$$X(f)\sum_{l=0}^{q} b(l)e^{-2\pi jflT} = Y(f)\sum_{l=0}^{p} a(l)e^{-2\pi jflT},$$

so the transfer function is

$$H(f) = \frac{Y(f)}{X(f)} = \frac{\sum_{l=0}^{q} b(l)e^{-2\pi jflT}}{\sum_{l=0}^{p} a(l)e^{-2\pi jflT}}. \qquad (6.3.12)$$

Example 6.3.2. A solution of the stochastic difference equation. This example was considered in Chap. 4, but let us observe that it is a special case of Example 6.3.1, with parameters $p = 1$, $q = 0$, and the input signal being the standard discrete white noise $W(n)$ with $\sigma_W^2 = 1$. In other words,

$$Y(n) = -a_1 Y(n-1) + b_0 W(n).$$

In view of (6.3.12), the power transfer function is

$$|H(f)|^2 = \frac{b_0}{1 + a_1 e^{-2\pi jf}} \cdot \frac{b_0}{1 + a_1 s e^{2\pi jf}} = \frac{b_0^2}{1 + a_1^2 + 2a_1 \cos 2\pi f},$$

with, again, all the relevant information contained in the frequency interval $-1/2 < f < 1/2$.

Given that the input is the standard white noise, we have that

$$S_Y(f) = |H(f)|^2 \cdot 1 = \frac{b_0^2}{1 + a_1^2 + 2a_1 \cos 2\pi f}. \qquad (6.3.13)$$

One way to find the output autocovariance sequence $\gamma_Y(n)$ would be to take into account the relationship (6.3.10) and expand (6.3.13) into the Fourier series; its coefficients will form the desired autocovariance sequence. This procedure is straightforward and requires only an application of the formula for the sum of a geometric series (see Sect. 6.4).

However, we would like to explore a different route here and employ a recursive procedure to find the output autocovariance sequence. First, observe that

$$\begin{aligned}
\gamma_Y(k) &= \mathbf{E}(Y(n)Y(n+k)) \\
&= \mathbf{E}(-a_1 Y(n-1) + b_0 X(n)) \cdot (-a_1 Y(n+k-1) + b_0 X(n+k)) \\
&= a_1^2 \mathbf{E}(Y(n-1)Y(n+k-1)) - a_1 b_0 \mathbf{E}(Y(n-1)X(n+k)) \\
&\quad - a_1 b_0 \mathbf{E}(X(n)Y(n+k-1)) + b_0^2 \mathbf{E}(X(n)X(n+k))
\end{aligned}$$

$$= a_1^2 \gamma_Y(k) - a_1 b_0 \gamma_{XY}(k-1) + b_0^2 \gamma_X(k),$$

where

$$\gamma_{XY}(k) = \mathbf{E}(X(n)Y(n+k))$$

is the cross-covariance sequence of signals $X(n)$ and $Y(n)$. So

$$\gamma_Y(k) = \frac{b_0}{1 - a_1^2}\left(-a_1\gamma_{XY}(k-1) + b_0\gamma_X(k)\right).$$

For $k = 0$,

$$\gamma_Y(0) = \sigma_Y^2 = \frac{b_0}{1 - a_1^2}\left(-a_1\mathbf{E}(X(n)Y(n-1)) + b_0\gamma_X(0)\right)$$

$$= \frac{b_0^2}{1 - a_1^2}\gamma_X(0) = \frac{b_0^2}{1 - a_1^2}.$$

For $k = 1$,

$$\gamma_Y(1) = \frac{b_0}{1 - a_1^2}\left(-a_1\gamma_{XY}(0) + b_0\gamma_X(1)\right) = \frac{b_0(-a_1)}{1 - a_1^2}\mathbf{E}(X(0)Y(0))$$

$$= \frac{b_0(-a_1)}{1 - a_1^2}\mathbf{E}\left(X(0)\left(a_1 Y(-1) + b_0 X(0)\right)\right) = \frac{b_0^2(-a_1)}{1 - a_1^2}.$$

For a general $k > 1$,

$$\gamma_Y(k) = \frac{b_0}{1 - a_1^2}\left(-a_1\gamma_{XY}(k-1) + b_0\gamma_X(k)\right),$$

and, as above,

$$\gamma_{XY}(k-1) = \mathbf{E}\left(X(0)Y(k-1)\right)$$

$$= \mathbf{E}\left(X(0)(-a_1 Y(k-2) + b_0 X(k-1))\right)$$

$$= (-a_1)\mathbf{E}\left(X(0)Y(k-2)\right)$$

$$= (-a_1)\gamma_{XY}(k-2) = \cdots = (-a_1)^{k-1}\gamma_{XY}(0) = b(0)(-a_1)^{k-1}.$$

Since the autocovariance sequence must be an even function of the variable k, we finally get, for any $k = \ldots, -2, -1, 0, 1, 2, \ldots$,

$$\gamma_Y(k) = \frac{b_0^2}{1 - a_1^2}(-a_1)^{|k|},$$

thus recovering the result from Chap. 4.

6.4 Problems and Exercises

In Exercises 6.4.1–6.4.3, also try solving the problem by first finding the autocovariance function of the output to see how hard the problem is in the time-domain
framework.

6.4.1. The impulse response function of a linear system is $h(t) = 1-t$ for $0 \leq t \leq 1$
and 0 elsewhere:

(a) Produce a graph of $h(t)$.
(b) Assume that the input is the standard white noise. Find the autocovariance function of the output.
(c) Find the power transfer function of the system, its equivalent-noise bandwidth,
and its half-power bandwidth.
(d) Assume that the input has the autocovariance function $\gamma_X(t) = 3/(1 + 4t^2)$.
Find the power spectrum of the output signal.
(e) Assume that the input has the autocovariance function $\gamma_X(t) = \exp(-4|t|)$.
Find the power spectrum of the output signal.
(f) Assume that the input has the autocovariance function $\gamma_X(t) = 1 - |t|$ for
$|t| < 1$ and 0 elsewhere. Find the power spectrum of the output signal.

6.4.2. The impulse response function of a linear system is $h(t) = e^{-2t}$ for $0 \leq t \leq$
2 and 0 elsewhere:

(a) Produce a graph of $h(t)$.
(b) Assume that the input is the standard white noise. Find the autocovariance function of the output.
(c) Find the power transfer function of the system, its equivalent-noise bandwidth,
and its half-power bandwidth.
(d) Assume that the input has the autocovariance function $\gamma_X(t) = 3/(1 + 4t^2)$.
Find the power spectrum of the output signal.
(e) Assume that the input has the autocovariance function $\gamma_X(t) = \exp(-4|t|)$.
Find the power spectrum of the output signal.
(f) Assume that the input has the autocovariance function $\gamma_X(t) = 1 - |t|$ for
$|t| < 1$ and 0 elsewhere. Find the power spectrum of the output signal.

6.4.3. The impulse response function of a linear system is $h(t) = e^{-0.05t}$ for $t \geq 10$
and 0 elsewhere:

(a) Produce a graph of $h(t)$.
(b) Assume that the input is the standard white noise. Find the autocovariance function of the output.
(c) Find the power transfer function of the system, its equivalent-noise bandwidth,
and its half-power bandwidth.
(d) Assume that the input has the autocovariance function $\gamma_X(t) = 3/(1 + 4t^2)$.
Find the power spectrum of the output signal.

(e) Assume that the input has the autocovariance function $\gamma_X(t) = \exp(-4|t|)$. Find the power spectrum of the output signal.

(f) Assume that the input has the autocovariance function $\gamma_X(t) = 1 - |t|$ for $|t| < 1$ and 0 elsewhere. Find the power spectrum of the output signal.

6.4.4. For a pair of random signals, $X(t)$ and $Y(t)$, the *cross-covariance*, γ_{XY}, is defined as follows:

$$\gamma_{XY}(t, s) = \mathbf{E}((X(t) - \mu_X(t))(Y(s) - \mu_Y(s))).$$

The random signals $X(t)$ and $Y(t)$ are said to be *jointly stationary* if they are stationary and their cross-covariance satisfies the condition

$$\gamma_{XY}(t, t + \tau) = \gamma_{XY}(\tau).$$

Consider the random signals

$$X(t) = a \cos(2\pi(f_0 t + \Theta)), \qquad Y(t) = b \sin(2\pi(f_0 t + \Theta)),$$

where a and b are nonrandom constants and Θ is uniformly distributed on $[0, 1]$. Find the cross-covariance function for X and Y. Are these signals jointly stationary?

6.4.5. Consider the circuit shown in Fig. 6.4.1.

Assume that the input, $X(t)$, is the standard white noise:

(a) Find the power spectra $S_Y(f)$ and $S_Z(f)$ of the outputs $Y(t)$ and $Z(t)$.
(b) Find the cross-covariance,

$$\gamma_{YZ}(\tau) = \mathbf{E}\Big(Z(t)Y(t + \tau)\Big),$$

between those two outputs.

6.4.6. Find the output autocovariance sequence for the discrete-time system representing a stochastic difference equation described in Example 6.3.2. Use the Fourier series expansion of formula (6.3.12).

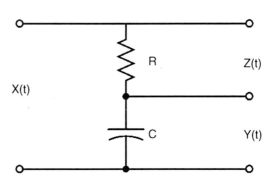

Fig. 6.4.1 The circuit discussed in Problem 6.4.5

Fig. 6.4.2 The circuit
discussed in Problem 6.4.7

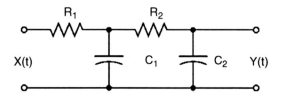

6.4.7. Consider the circuit shown in Fig. 6.4.2:

(a) Assume that the input is the standard white noise. Find the power spectrum
$S_Y(f)$ and the autocovariance function $\gamma_Y(\tau)$ of the output $Y(t)$. *Hint:* Think
about the above circuit as two simple RC filters in series.
(b) Find the half-power and equivalent-noise bandwidth for the system shown in
Fig. 6.4.2 in the case when $R_1 = R_2$ and $C_1 = C_2$.

6.4.8. Show that a *continuum limit of RC filters in series* has a Gaussian p.d.f.-like
power transfer function. Then prove that a white noise transmitted through such a
filter yields a stationary signal on the output with a Gaussian p.d.f.-like autocovariance function. More precisely, consider n rescaled RC filters in series, each with
the time constant equal to $RC/\sqrt{n}$. Calculate its power transfer function, and take
$n \to \infty$, to obtain the sought power transfer function of the form

$$|H(f)|^2 = e^{-(2\pi RCf)^2}.$$

Hint: Use the basic calculus fact that $(1 + 1/x)^x \to e$, as $x \to \infty$. Then use
the inverse Fourier transform to calculate the desired ACvF of the output. Note the
following fundamental fact: The (inverse) Fourier transform of a Gaussian p.d.f.-like
function is also a Gaussian p.d.f.-like function.

Chapter 7
Optimization of Signal-to-Noise Ratio in Linear Systems

Useful, deterministic signals passing through various transmission devices often acquire extraneous random components due to, say, thermal noise in conducting materials, radio clutter or *aurora borealis* magnetic field fluctuations in the atmosphere, or deliberate jamming in warfare. If there exists some prior information about the nature of the original useful signal and the contaminating random noise, it is possible to devise algorithms to improve the relative power of the useful component of the signal, or, in other words, to increase the *signal-to-noise ratio* of the signal, by passing it through a filter designed for the purpose. In this short chapter we give a few examples of such designs just to show how the previously introduced techniques of analysis of random signals can be applied in this context.

7.1 Parametric Optimization for a Fixed Filter Structure

The general problem of optimization (maximization) of the signal-to-noise ratio in a linear system schematically pictured here,

$$x(t) + N(t) \longrightarrow \boxed{h(t)} \longrightarrow y(t) + M(t),$$

can be formulated as follows: Consider a linear filter (system) characterized by its impulse response function $h(t)$ with the input signal $X(t)$ of the form

$$X(t) = x(t) + N(t), \tag{7.1.1}$$

where $x(t)$ is a deterministic "useful" signal, and $N(t)$ is a random stationary "noise" signal with zero mean and autocovariance function $\gamma_N(t)$. Given the linearity of the system, the output signal $Y(t)$ is of the form

$$Y(t) = y(t) + M(t), \tag{7.1.2}$$

W.A. Woyczyński, *A First Course in Statistics for Signal Analysis*,
DOI 10.1007/978-0-8176-8101-2_7, © Springer Science+Business Media, LLC 2011

where the deterministic "useful" output component is

$$y(t) = \int_{-\infty}^{\infty} x(s)h(t-s)ds, \tag{7.1.3}$$

and the "noise" output is a stationary zero-mean signal with the autocovariance function

$$\gamma_M(\tau) = \int_{-\infty}^{\infty}\int_{-\infty}^{\infty} \gamma_N(\tau - s + u)h(s)h(u)\,ds\,du.$$

The task is as follows: Given the shape of the input signal, design the structure of the filter which would maximize the signal-to-noise power ratio on the output. More precisely, we need to find an impulse response function $h(t)$ such that, for a given detection time t, the signal-to-noise ratio

$$S/N = \frac{PW_y(t)}{E(PW_M)} \tag{7.1.4}$$

is maximized over all possible impulse response functions; in brief, we want to find $h(t)$ for which

$$S/N = \text{max}.$$

Here, $PW_y(t) = y^2(t)$ is the instantaneous power of the output signal, and $E(PW_M) = \gamma_M(0) = \sigma_M^2$ is the mean power of the output noise. Hence, the optimization problem is to find $h(t)$, and also the detection time t_0, such that

$$S/N = \frac{y^2(t_0)}{\gamma_M(0)} = \frac{y^2(t_0)}{\sigma_M^2} = \text{max}. \tag{7.1.5}$$

In the present section we will take a look at a relatively simple situation when the general structure of the filter is essentially fixed and only certain parameters, including the detection time t_0, need to be optimized.

To show the essence of our approach, we will just consider the RC filter with the impulse response function

$$h(t) = be^{-bt} \cdot u(t), \tag{7.1.6}$$

with a single parameter $b = 1/RC$ to be determined in addition to the optimal detection time t_0.

Suppose that the "useful" input signal we are trying to detect on the output is a rectangular impulse

$$x(t) = \begin{cases} A, & \text{for} \quad 0 \le t \le T; \\ 0, & \text{elsewhere}, \end{cases} \tag{7.1.7}$$

and that the input noise is a white noise of "amplitude" N_0, with the autocovariance $\gamma_N(t) = N_0\delta(t)$.

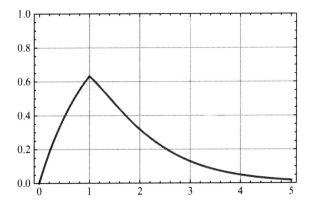

Fig. 7.1.1 Response $y(t)$ (7.1.8) of the RC filter (7.1.6) to the rectangular input signal $x(t)$ (7.1.7). The parameter values are $T = 1$, $A = 1$, and $b = 1/RC = 1$. The maximum is clearly attained for $t_0 = T$

The deterministic "useful" output signal is

$$
\begin{aligned}
y(t) &= \int_{-\infty}^{\infty} x(s)h(-(s-t))\,ds \\
&= \begin{cases} \int_0^t Abe^{-b(t-s)}\,ds, & \text{for } 0 < t < T; \\ \int_0^T Abe^{-b(t-s)}\,ds, & \text{for } t \geq T, \end{cases} \\
&= \begin{cases} A(1 - e^{-bt}), & \text{for } 0 < t \leq T; \\ A(1 - e^{-bT})e^{-b(t-T)}, & \text{for } t \geq T, \end{cases}
\end{aligned} \tag{7.1.8}
$$

and is pictured in Fig. 7.1.1.

Clearly, the maximum of the output signal is attained at $t_0 = T$. On the other hand, as calculated in Chap. 6, the autocovariance function of the output noise is

$$
\gamma_M(\tau) = N_0 \frac{b}{2} e^{-b\tau},
$$

so that, at the already-optimized detection time $t_0 = T$,

$$
\frac{S}{N} = \frac{y^2(T)}{\gamma_M(0)} = \frac{A^2[1 - e^{-bT}]^2}{bN_0/2}.
$$

To simplify our calculations, we will substitute $z = bT$. Now, our final task is to find the maximum of the function

$$
\frac{S}{N}(z) = \frac{2A^2 T}{N_0} \cdot \frac{(1 - e^{-z})^2}{z} \tag{7.1.9}
$$

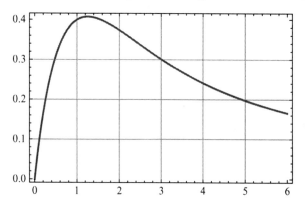

Fig. 7.1.2 Graph of the factor $(1 - e^{-z})^2/z$ in formula (7.1.9) for the signal-to-noise ratio $\mathcal{S}/\mathcal{N}(z)$

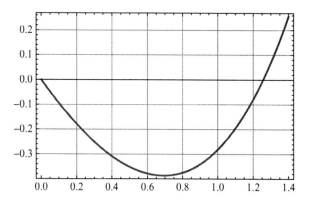

Fig. 7.1.3 A plot of the function $e^z - 1 - 2z = 0$. The nontrivial zero is approximately at $z_{max} = 1.25$

of one variable z. The function $\mathcal{S}/\mathcal{N}(z)$, although simple-looking, is a little tricky, and we will start the exploration of its maximum by graphing it; see Fig. 7.1.2. To find the location of the maximum, we calculate the derivative and try to solve the equation

$$\frac{d}{dz}\frac{(1 - e^{-z})^2}{z} = \frac{2(1 - e^{-z})e^{-z}z - (1 - e^{-z})^2}{z^2} = 0.$$

Although the above equation can be easily simplified to the equation

$$e^z - 1 - 2z = 0,$$

the latter cannot be solved explicitly. So, as usual, as the first step we explore the solution graphically; see Fig. 7.1.3. The nontrivial zero is approximately at $z_{max} = 1.25$, which gives $b_{max} = 1.25/T$, so that the optimal RC constant is

$$RC_{\max} \approx \frac{1}{b_{\max}} = \frac{T}{1.25} = 0.8T. \tag{7.1.10}$$

Note that $RC_{\max}$ is independent of the "amplitude," N_0, of the input noise.

Evaluated at the optimal values of parameters t_0 and b, the maximum available signal-to-noise ratio is

$$\frac{S}{N}\bigg|_{\max} \approx \frac{y^2(T)}{b_{\max} N_0/2} = \frac{2A^2[1 - e^{-b_{\max}T}]^2}{b_{\max} N_0} = 0.81 \cdot \frac{A^2 T}{N_0}. \tag{7.1.11}$$

It is proportional to the signal's duration T, and to the square of its amplitude A, but is inversely proportional to the "amplitude," N_0, of the noise.

7.2 Filter Structure Matched to Input Signal

In this section we will solve a more ambitious problem of designing the structure of the filter to maximize the signal-to-noise ratio on the output rather than just optimizing filter parameters. To be more precise, the task at hand is to find an impulse response function $h(t)$, and the detection time t_0, such that

$$S/N = \frac{y^2(t_0)}{\sigma_M^2} = \max, \tag{7.2.1}$$

for a given deterministic (nonrandom) input signal $x(t)$ transmitted in the presence of the white noise input $N(t)$ with autocovariance function $\gamma_N(t) = N_0\delta(t)$, where, as before, $x(t) = 0$, for $t \leq 0$, and

$$y(t) = \int_0^\infty x(t - s)h(s)\, ds. \tag{7.2.2}$$

For the output noise,

$$\sigma_M^2 = \gamma_M(0) = \int_0^\infty \left(\int_0^\infty \delta(u - s)h(u)\, du \right) h(s)\, ds = N_0 \int_0^\infty h^2(s)\, ds. \tag{7.2.3}$$

In this situation

$$S/N = \frac{y^2(t_0)}{\sigma_M^2} = \frac{(\int_0^\infty x(t_0 - s)h(s)\, ds)^2}{N_0 \int_0^\infty h^2(s)\, ds}. \tag{7.2.4}$$

In view of the Cauchy–Schwartz inequality,

$$S/\mathcal{N} \le \frac{\int_0^\infty x^2(t_0 - s)\,ds \cdot \int_0^\infty h^2(s)\,ds}{N_0 \int_0^\infty h^2(s)\,ds} = \frac{1}{N_0} \int_0^\infty x^2(t_0 - s)\,ds, \qquad (7.2.5)$$

with the equality, that is, the maximum for $S/\mathcal{N}$, achieved when the two factors, $h(s)$ and $x(t_0 - s)$, in the scalar product in the numerator of (7.2.4) are linearly dependent. In other words, for any constant c, the impulse response function

$$h(s) = cx(t_0 - s)u(s) = cx(-(s - t_0))u(s) \qquad (7.2.6)$$

gives the optimal structure of the filter and maximizes the $S/\mathcal{N}$ ratio. This so-called *matching filter* has the impulse response function equal to the input signal $x(t)$ run backward in time, then shifted to the right by t_0, and, finally, cut off at 0.

With the selection of the matching filter, in view of (7.2.4), the maximal value of the $S/\mathcal{N}$ ratio is

$$S/\mathcal{N}_{\max} = \frac{(\int_0^\infty x(t_0 - s)cx(t_0 - s)u(s)\,ds)^2}{N_0 \int_0^\infty (cx(t_0 - s)u(s))^2\,ds} = \frac{\int_0^\infty x^2(t_0 - s)\,ds}{N_0}. \qquad (7.2.7)$$

Example 7.2.1 (Matching filter for a rectangular input signal). Consider a rectangular input signal of the form

$$x(t) = \begin{cases} A, & \text{for } 0 < t < T; \\ 0, & \text{elsewhere}, \end{cases}$$

transmitted in the presence of an additive white noise with autocovariance function $\gamma_N(t) = N_0\delta(t)$. According to formula (7.2.6), its matching filter at detection time t_0 is

$$h(t) = \begin{cases} A, & \text{for } 0 < t < t_0; \\ 0, & \text{elsewhere}, \end{cases}$$

if $0 \le t_0 \le T$, and

$$h(t) = \begin{cases} A, & \text{for } t_0 - T < t < t_0; \\ 0, & \text{elsewhere}, \end{cases}$$

if $t_0 > T$. So the $S/\mathcal{N}_{\max}$, as a function of the detection time t_0, is

$$S/\mathcal{N}_{\max}(t_0) = \begin{cases} A^2 t_0/N_0, & \text{for } 0 < t_0 < T; \\ A^2 T/N_0, & \text{for } t_0 > T. \end{cases}$$

Clearly, the earliest detection time t_0 to maximize $S/\mathcal{N}_{\max}(t_0)$ is $t_0 = T$ (see Fig. 7.2.1).

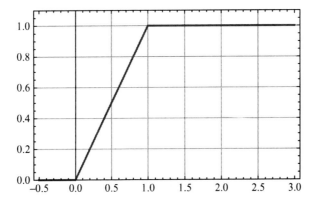

Fig. 7.2.1 The dependence of the optimal signal-to-noise ratio on the detection time t_0 for the matching filter from Example 7.2.1. The input signal is the sum of a rectangular signal of amplitude $A = 1$, and duration $T = 1$, and the white noise with autocovariance function $\gamma_N(t) = \delta(t)$

At the optimal detection time $t_0 = T$, or any later detection time,

$$S/N_{\max} = \frac{A^2 T}{N_0}. \tag{7.2.8}$$

This result should be compared with the maximum signal-to-noise ratio $0.81 A^2 T/N_0$ [see (7.1.11)] obtained in Sect. 7.1 by optimally tuning the RC-filter: The best matching filter gives about a 25% gain in the signal-to-noise ratio over the best RC-filter.

It is also instructive to trace the behavior of the deterministic part $y(t)$ of the output signal for the matching filter as a function of the detection time t_0. Formula (7.2.2) applied to the matching filter immediately gives that, for $0 < t_0 < T$,

$$y(t) = \begin{cases} A^2 t, & \text{for } 0 < t < t_0; \\ A^2 t_0, & \text{for } t_0 < t < T; \\ -A^2(t - (t_0 + T)), & \text{for } T < t < t_0 + T; \\ 0, & \text{elsewhere}, \end{cases} \tag{7.2.9}$$

and, for $t_0 \geq T$,

$$y(t) = \begin{cases} A^2(t - (t_0 - T)), & \text{for } t_0 - T < t < t_0; \\ -A^2(t - (t_0 + T)), & \text{for } t_0 < t < t_0 + T; \\ 0, & \text{elsewhere}. \end{cases} \tag{7.2.10}$$

These two output signals are depicted in Fig. 7.2.2.

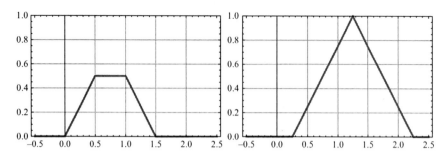

Fig. 7.2.2 The response $y(t)$ of the matching filter for the rectangular input signal with amplitude $A = 1$ and duration $T = 1$ (see Example 7.2.1). *Top:* For detection time $t_0 = 0.25 < T = 1$. *Bottom:* For detection time $t_0 = 1.25 > T = 1$

7.3 The Wiener Filter

Acausal filter. Given stationary random signals $X(t)$, and $Y(t)$, the problem is to find a (not necessarily causal) impulse response function $h(t)$ such that the mean-square distance between $Y(t)$ and the output signal,

$$Y_h(t) = \int_{-\infty}^{\infty} X(t - s)h(s)\,ds,$$

is the smallest possible. In other words, we need $h(t)$ minimizing the error quantity

$$\mathbf{E}\left(Y(t) - Y_h(t)\right)^2.$$

In the space of all finite variance (always zero-mean) random quantities equipped with the covariance as the scalar product, the best approximation $Y_h(t)$ of a random quantity $Y(t)$ by elements of the linear subspace $\mathcal{X}$ spanned by linear combinations of values of $X(t - s), -\infty < s < \infty$, is given by the orthogonal projection of $X(t)$ on $\mathcal{X}$.[1] That means that the difference $Y(t) - Y_h(t)$ must be orthogonal to all $X(t - s), -\infty < s < \infty$, or more formally,

$$\mathbf{E}\left((Y(t) - Y_h(t)) \cdot X(t - s)\right) = \mathbf{E}\left(Y(t) \cdot X(t - s)\right)$$
$$-\mathbf{E}\left(\int_{-\infty}^{\infty} X(t - u)h(u)\,du \cdot X(t - s)\right)$$
$$= \gamma_{YX}(s) - \int_{-\infty}^{\infty} \gamma_X(s - u)h(u)\,du = 0,$$

---

[1] This argument is analogous to the one encountered in Chap. 2, when we discussed the best approximation in power of deterministic periodic signals by their Fourier series.

for all s, $-\infty < s < \infty$. Hence, the optimal $h(t)$ can be found by solving, for each s, the integral equation

$$\gamma_{YX}(s) = \int_{-\infty}^{\infty} \gamma_X(s - u)h(u)\,du, \qquad (7.3.1)$$

which involves only the autocovariance function $\gamma_X(s)$ and the cross-correlation function $\gamma_{YX}(s)$. The solution is readily found in the frequency domain. Remembering that the Fourier transform of a convolution is the product of Fourier transforms, and denoting by $H(f)$ the transfer function (the Fourier transform of the impulse response function) of the optimal $h(t)$, (7.3.1) can be rewritten in the form

$$S_{YX}(f) = S_X(f) \cdot H(f),$$

which immediately gives the explicit formula for the transfer function of the optimal filter:

$$H(f) = \frac{S_{YX}(f)}{S_X(f)}. \qquad (7.3.2)$$

The minimal error can then also be calculated explicitly:

$$\mathbf{E}\Big(Y(t) - Y_h(t)\Big)^2 = \gamma_Y(0) - \int_{-\infty}^{\infty} \gamma_{YX}(s)h(s)\,ds, \qquad (7.3.3)$$

or, in terms of the optimal transfer function, using Parseval's formula for the last integral, we have

$$\mathbf{E}\Big(Y(t) - Y_h(t)\Big)^2 = \int_{-\infty}^{\infty} \Big(S_Y(f) - S_{YX}^*(f)H(f)\Big)\,df. \qquad (7.3.4)$$

Example 7.3.1 (Filtering white noise out of a stationary signal). Assume that the signal $X(t)$ is the sum of a "useful" signal $Y(t)$ and noise $N(t)$, that is, $X(t) = Y(t) + N(t)$, where $Y(t)$ has the power spectrum

$$S_Y(f) = \frac{1}{1 + f^2},$$

and is uncorrelated with the white noise $N(t)$, which is assumed to have the power spectrum $S_N(f) \equiv 1$. Then

$$S_{YX}(f) = S_Y(f) = \frac{1}{1 + f^2} \quad \text{and} \quad S_X(f) = S_Y(f) + S_N(f) = \frac{2 + f^2}{1 + f^2}.$$

The transfer function of the optimal filter is then

$$H(f) = \frac{S_{YX}(f)}{S_X(f)} = \frac{1}{2 + f^2},$$

with the corresponding impulse response function

$$h(t) = \frac{1}{2\sqrt{2}} e^{-\sqrt{2}|t|},$$

and the error is

$$\mathbf{E}\left(Y(t) - Y_h(t)\right)^2 = \int_{-\infty}^{\infty} \left(\frac{1}{1+f^2} - \frac{1}{1+f^2} \cdot \frac{1}{2+f^2}\right) df$$

$$= \int_{-\infty}^{\infty} \frac{1}{2+f^2} df = \frac{\pi}{\sqrt{2}}.$$

Causal filter. For given stationary random signals $X(t)$ and $Y(t)$, the construction of the optimal *causal filter* requires finding a causal impulse response function $h(t) = 0$, for $t \leq 0$, such that the error

$$\mathbf{E}\left(Y(t) - \int_0^{\infty} X(t-s)h(s)\, ds\right)^2$$

is minimal. In other words, we are trying to find the best mean-square approximation to $Y(t)$ by (continuous) linear combinations of the *past* values of $X(t)$. Using the same orthogonality argument we applied for the acausal optimal filter, we obtain another integral equation for the optimal $h(t)$:

$$\gamma_{YX}(s) = \int_0^{\infty} \gamma_X(s-u)h(u)\, du,$$

this time valid only for all $s > 0$. This equation is traditionally called the *Wiener–Hopf equation*. It is clear that to solve the above equation via an integral transform method, we have to replace the Fourier transform used in the acausal case by the Laplace transform. However, the details here are more involved and, for the solution, we refer the reader to the literature on the subject.[2]

7.4 Problems and Exercises

7.4.1. The triangular signal $x(t) = 0.01t$, for $0 < t < 0.01$, and 0 elsewhere, is combined with white noise having a flat power spectrum of $2\,\mathrm{V}^2/\mathrm{Hz}$. Find the value of the RC constant such that the signal-to-noise ratio at the output of the RC filter is maximal at $t = 0.01$ s.

---

[2] Norbert Wiener's original *Extrapolation, Interpolation, and Smoothing of Stationary Time Series*, MIT Press and Wiley, New York, 1950, is still very readable, but also see Chap. 10 of A. Papoulis, *Signal Analysis*, McGraw-Hill, New York, 1977.

7.4.2. A signal of the form $x(t) = 5e^{-(t+2)}u(t)$ is to be detected in the presence of white noise with a flat power spectrum of $0.25 \text{ V}^2/\text{Hz}$ using a matched filter.

(a) For $t_0 = 2$, find the value of the impulse response of the matched filter at $t = 0, 2, 4$.
(b) Find the maximum output signal-to-noise ratio that can be achieved if $t_0 = \infty$.
(c) Find the detection time t_0 that should be used to achieve an output signal-to-noise ratio that is equal to 95% of the maximum signal-to-noise ratio discovered in part (b).
(d) The signal $x(t) = 5e^{-(t+2)}u(t)$ is combined with white noise having a power spectrum of $2 \text{ V}^2/\text{Hz}$. Find the value of RC such that the signal-to-noise ratio at the output of the RC filter is maximal at $t = 0.01$ s.

7.4.3. Repeat construction of the optimal filter from Example 7.3.1 in the case when the useful signal $Y(t)$ has a more general power spectrum

$$S_Y(f) = \frac{a}{b^2 + f^2},$$

and the uncorrelated white noise $N(t)$ has the arbitrary power spectrum $S_N(f) \equiv \mathcal{N}$. Discuss the properties of this filter when the noise power is much bigger than the power of the useful signal, that is, when $\mathcal{N} \gg S_Y(f)$. Construct the optimal acausal filters for other selected spectra of $Y(t)$ and $N(t)$.

Chapter 8
Gaussian Signals, Covariance Matrices, and Sample Path Properties

In general, determining the shape of the sample paths of a random signal $X(t)$ requires knowledge of n-D (or, in the terminology of signal processing, n-point) probabilities

$$\mathbf{P}\Big(a_1 < X(t_1) < b_1, \ldots, a_n < X(t_n) < b_n\Big),$$

for an arbitrary n and arbitrary windows $a_1 < b_1, \ldots, a_n < b_n$. But, usually, this information cannot be recovered if the only signal characteristic known is the autocorrelation function. The latter depends on the two-point distributions but does not uniquely determine them. However, in the case of Gaussian signals, the autocovariances determine not only two-point probability distributions but also all the n-point probability distributions, so that complete information is available within the second-order theory. In particular, that means that you only have to estimate means and covariances to obtain the complete model. Also, in the Gaussian universe, the weak stationarity implies the strict stationarity as defined in Chap. 4. For the sake of simplicity, all signals in this chapter are assumed to be real-valued. The chapter ends with a more subtle analysis of sample path properties of stationary signals such as continuity and differentiability; in the Gaussian case these issues have fairly complete answers.

Of course, faced with real-world data, the proposition that they are distributed according to a Gaussian distribution must be tested rigorously. Many such tests have been developed by statisticians.[1] In other cases, one can make an argument in favor of such a hypothesis based on the central limit theorem (3.5.5) and (3.5.6).

8.1 Linear Transformations of Random Vectors

In Chap. 3 we calculated probability distributions of transformed random quantities. Repeating that procedure in the case of a linear transformation of the 1D random quantity X given by the formula

---

[1] See, e.g., M. Denker and W. A. Woyczyński's book mentioned in previous chapters.

W.A. Woyczyński, *A First Course in Statistics for Signal Analysis*,
DOI 10.1007/978-0-8176-8101-2_8, © Springer Science+Business Media, LLC 2011

$$Y = aX, \qquad a > 0, \tag{8.1.1}$$

we can obtain the cumulative distribution function (c.d.f.) $F_Y(y)$ of the random quantity Y in terms of the c.d.f. $F_X(x)$ of the random quantity X as follows:

$$F_Y(y) = P(Y \leq y) = P(aX \leq y) = P(X \leq y/a) = F_X(y/a). \tag{8.1.2}$$

To obtain an analogous formula for the probability density functions (p.d.f.s), it suffices to differentiate both sides of (8.1.2) to see that

$$f_Y(y) = \frac{d}{dy} F_Y(y) = \frac{1}{a} f_X\left(\frac{y}{a}\right). \tag{8.1.3}$$

Example 8.1.1. Consider a standard 1D Gaussian random quantity $X \sim N(0, 1)$ with the p.d.f.

$$f_X(x) = \frac{1}{\sqrt{2\pi}} e^{-x^2/2}. \tag{8.1.4}$$

Then the random quantity $Y = aX$, $a > 0$, has the p.d.f.

$$f_Y(y) = \frac{1}{\sqrt{2\pi}a} e^{-\frac{y^2}{2a^2}}. \tag{8.1.5}$$

Obviously, the expectation is

$$EY = E(aX) = aEX = 0,$$

and the variance of Y is

$$\sigma_Y^2 = E(aX)^2 = a^2 EX^2 = a^2. \tag{8.1.6}$$

If we conduct the same argument for $a < 0$, the p.d.f. of $Y = aX$ will be

$$f_Y(y) = \frac{1}{\sqrt{2\pi}(-a)} e^{-\frac{x^2}{2a^2}}. \tag{8.1.7}$$

Thus formulas (8.1.6) and (8.1.7) can be unified in a single statement: If $X \sim N(0, 1)$, then, for any $a \neq 0$, the random quantity $Y = aX$ has the p.d.f.

$$f_Y(y) = \frac{1}{\sqrt{2\pi}|a|} e^{-\frac{x^2}{2a^2}}. \tag{8.1.8}$$

Using the above elementary reasoning as a model, we will now derive the formula for a d-dimensional p.d.f.

$$f_{\vec{Y}}(\vec{y}) = f_{\vec{Y}}(y_1, \ldots, y_d)$$

of a random (column) vector

$$\vec{Y} = \begin{pmatrix} Y_1 \\ \vdots \\ Y_d \end{pmatrix}$$

obtained by a nondegenerate linear transformation

$$\vec{Y} = \mathbf{A}\vec{X} \tag{8.1.9}$$

consisting of multiplication of the random vector

$$\vec{X} = \begin{pmatrix} X_1 \\ \vdots \\ X_d \end{pmatrix},$$

with a known p.d.f.

$$f_{\vec{X}}(\vec{x}) = f_{\vec{X}}(x_1, \ldots, x_d),$$

by a fixed nondegenerate nonrandom matrix

$$\mathbf{A} = \begin{pmatrix} a_{11}, \ldots, a_{1d} \\ \cdots \\ a_{d1}, \ldots, a_{dd} \end{pmatrix}.$$

In other words, we assume that $\det(A) \neq 0$, or, equivalently, that the rows of the matrix A form a linearly independent system of vectors.

In terms of its coordinates, the result of the linear transformation (8.1.9) can be written in the explicit form

$$\vec{Y} = \begin{pmatrix} a_{11}X_1 + a_{12}X_2 + \cdots + a_{1d}X_d \\ a_{21}X_1 + a_{22}X_2 + \cdots + a_{2d}X_d \\ \cdots \quad \cdots \quad \cdots \quad \cdots \\ a_{d1}X_1 + a_{d2}X_2 + \cdots + a_{dd}X_d \end{pmatrix}.$$

To calculate the probability distribution of $\vec{Y}$ following the above 1D approach, we must make use of the essential assumption of invertibility of the matrix $\mathbf{A}$, an analog of the assumption $a \neq 0$ in the 1D case. Then, for a domain D in the d-dimensional space $\mathbf{R}^d$,

$$\mathbf{P}(\vec{Y} \in D) = \mathbf{P}(\mathbf{A}\vec{X} \in D) = \mathbf{P}(\vec{X} \in \mathbf{A}^{-1}D). \tag{8.1.10}$$

This identity can be rewritten in terms of the p.d.f.s of $\vec{Y}$ and $\vec{X}$ as follows:

$$\int_D f_{\vec{Y}}(\vec{y}) \, dy_1 \cdots \cdots \, dy_d = \int_{\mathbf{A}^{-1}D} f_{\vec{X}}(\vec{x}) \, dx_1 \cdots \cdots \, dx_d.$$

Making a substitution $\vec{x} = \mathbf{A}^{-1}\vec{z}$ in the second integral, in view of the d-dimensional change-of-variables formula, we get that

$$\int_D f_{\vec{Y}}(\vec{y})\, dy_1 \cdots \cdot dy_d = \int_D f_{\vec{X}}(\mathbf{A}^{-1}\vec{z}) \cdot |\det(\mathbf{A}^{-1})|\, dz_1 \cdots \cdot dz_d,$$

where $\det(\mathbf{A}^{-1})$ is just the Jacobian of the substitution $\vec{x} = \mathbf{A}^{-1}\vec{z}$. Remembering that the determinant of the inverse matrix $\mathbf{A}^{-1}$ is the reciprocal of the determinant of the matrix $\mathbf{A}$, we get the identity

$$\int_D f_{\vec{Y}}(\vec{y})\, dy_1 \cdots \cdot dy_d = \int_D \frac{f_{\vec{X}}(\mathbf{A}^{-1}\vec{z})}{|\det(\mathbf{A})|}\, dz_1 \cdots \cdot dz_d.$$

Since this identity holds true for any domain D, the integrands on both sides must be equal, which gives the final formula for the p.d.f. of $\vec{Y}$:

$$f_{\vec{Y}}(\vec{y}) = \frac{f_{\vec{X}}(\mathbf{A}^{-1}\vec{y})}{|\det(\mathbf{A})|}, \quad \text{if} \quad \det(\mathbf{A}) \neq 0. \tag{8.1.11}$$

The 1D formula (8.1.3) is, obviously, the special case of the above general result.

8.2 Gaussian Random Vectors

As in the one-dimensional case, all nondegenerate zero-mean d-dimensional Gaussian random vectors can be obtained as nondegenerate linear transformations of a standard d-D Gaussian random vector

$$\vec{X} = \begin{pmatrix} X_1 \\ \vdots \\ X_d \end{pmatrix}$$

in which the coordinates $X_1, \ldots, X_d$ are independent $N(0, 1)$ random quantities. Because of their independence, the d-dimensional p.d.f. of $\vec{X}$ is the product of 1D $N(0, 1)$ p.d.f.s and is thus of the product form

$$f_{\vec{X}}(\vec{x}) = \frac{e^{\frac{-x_1^2}{2}}}{\sqrt{2\pi}} \cdots \cdot \frac{e^{\frac{-x_d^2}{2}}}{\sqrt{2\pi}} = \frac{1}{(2\pi)^{d/2}} e^{-\frac{1}{2}(x_1^2 + \cdots + x_d^2)}$$

$$= \frac{1}{(2\pi)^{d/2}} e^{-\frac{1}{2}\|\vec{x}\|^2} = \frac{1}{(2\pi)^{d/2}} e^{-\frac{1}{2}\vec{x}^T \vec{x}}, \tag{8.2.1}$$

where $\|\vec{x}\|$ stands for the norm (magnitude) of the vector $\vec{x}$, and the superscript T denotes the transpose of a matrix. Indeed,

$$\vec{x}^T\vec{x} = (x_1,\ldots,x_d)\cdot \begin{pmatrix} x_1 \\ \vdots \\ x_d \end{pmatrix} = x_1^2 + \cdots + x_d^2 = \|\vec{x}\|^2.$$

It is the latter form in (8.2.1) that will be useful now in applying formula (8.1.11). Indeed, substituting the last expression for $f_{\vec{X}}(\vec{x})$ in (8.2.1) into (8.1.11), one immediately gets[2]

$$\begin{aligned} f_{\vec{Y}}(\vec{y}) &= \frac{1}{(2\pi)^{d/2}|\det(A)|} e^{-\frac{1}{2}\|A^{-1}\vec{y}\|^2} \\ &= \frac{1}{(2\pi)^{d/2}|\det(A)|} e^{-\frac{1}{2}(A^{-1}\vec{y})^T\cdot(A^{-1}\vec{y})} \\ &= \frac{1}{(2\pi)^{d/2}|\det(A)|} e^{-\frac{1}{2}\vec{y}^T(AA^T)^{-1}\vec{y}}. \end{aligned} \qquad (8.2.2)$$

Thus formula (8.2.2) gives the general form of the d-dimensional zero-mean Gaussian p.d.f., and just as we identified the parameter a^2 in the 1D case (8.1.5) and (8.1.6) as the variance of the random quantity Y, we can identify entries of the matrix

$$\Gamma = AA^T \qquad (8.2.3)$$

appearing in the exponent in (8.2.2) as statistically significant parameters of the random vector $\vec{Y}$.

To see what they are, let us first calculate the entries γ_{ij}, $i,j = 1,2,\ldots,d$, of the matrix Γ:

$$\gamma_{ij} = a_{i1}a_{j1} + a_{i2}a_{j2} + \cdots + a_{id}a_{jd}. \qquad (8.2.4)$$

On the other hand, covariances (we are working with zero-mean vectors!) of different components of the random vector $\vec{Y}$ are

$$\begin{aligned} E(Y_i Y_j) &= E\Big((a_{i1}X_1 + \cdots + a_{id}X_d)\cdot(a_{j1}X_1 + \cdots + a_{jd}X_d)\Big) \\ &= a_{i1}a_{j1} + a_{i2}a_{j2} + \cdots + a_{id}a_{jd} \end{aligned} \qquad (8.2.5)$$

because $EX_i X_j = 1$ if $i = j$ and $= 0$ if $i \neq j$.

So it turns out that

$$\Gamma = (\gamma_{ij}) = (EY_i Y_j), \qquad (8.2.6)$$

---

[2] Remember that for any matrices $\mathbf{M}$ and $\mathbf{N}$, we have $(\mathbf{MN})^T = \mathbf{N}^T\mathbf{M}^T$, $(\mathbf{MN})^{-1} = \mathbf{N}^{-1}\mathbf{M}^{-1}$, and $(\mathbf{M}^T)^{-1} = (\mathbf{M}^{-1})^T$.

and the matrix $\Gamma = (\gamma_{ij})$ is simply the covariance matrix of the general zero-mean Gaussian random vector $\vec{Y}$. Thus, since

$$\det(\Gamma) = \det(\mathbf{A}\mathbf{A}^T) = \det(\mathbf{A}) \cdot \det(\mathbf{A}^T) = (\det(\mathbf{A}))^2,$$

we finally get that the p.d.f. of $\vec{Y}$ can be written in the form

$$f_{\vec{Y}}(\vec{y}) = \frac{1}{(2\pi)^{d/2}|\det(\Gamma)|^{1/2}}e^{-\frac{1}{2}\vec{y}^T\Gamma^{-1}\vec{y}}, \tag{8.2.7}$$

where Γ is the covariance matrix of $\vec{Y}$ satisfying the nondegeneracy condition $\det(\Gamma) \neq 0$.

Remark 8.2.1 (Gaussian random vectors with nonzero mean). Of course, to get the p.d.f. of a general Gaussian random vector with nonzero expectation

$$\mathbf{E}\vec{Y} = \vec{\mu} = (\mu_1, \dots, \mu_d)^T,$$

it suffices to shift the p.d.f. (8.2.7) by $\vec{\mu}$ to obtain

$$f_{\vec{Y}}(\vec{y}) = \frac{1}{(2\pi)^{d/2}|\det(\Sigma)|^{1/2}}e^{-\frac{1}{2}(\vec{y}-\vec{\mu})^T\Sigma^{-1}(\vec{y}-\vec{\mu})}, \tag{8.2.8a}$$

where

$$\Sigma = (\sigma_{ij}) = (\mathbf{E}(Y_i - \mu_i)(Y_j - \mu_j)) \tag{8.2.8b}$$

is the *covariance matrix* of $\vec{Y}$. A Gaussian random vector with a joint p.d.f. given by formulas (8.2.7) and (8.2.8b) is often called a normal $N(\vec{\mu}, \Sigma)$ random vector.

Example 8.2.1 (2D zero-mean Gaussian random vectors (see, also, Example 3.3.2)). Let us carry out the above calculation explicitly in the special case of dimension $d = 2$. Then the covariance matrix

$$\Gamma = \begin{pmatrix} \mathbf{E}Y_1Y_1 & \mathbf{E}Y_1Y_2 \\ \mathbf{E}Y_2Y_1 & \mathbf{E}Y_2Y_2 \end{pmatrix} = \begin{pmatrix} \sigma_1^2 & \sigma_1\sigma_2\rho \\ \sigma_1\sigma_2\rho & \sigma_2^2 \end{pmatrix},$$

where the variances of coordinate vectors are

$$\sigma_1^2 = \mathbf{E}Y_1^2, \qquad \sigma_2^2 = \mathbf{E}Y_2^2,$$

and the correlation coefficient of the two components is

$$\rho = \frac{\mathbf{E}Y_1Y_2}{\sigma_1\sigma_2}.$$

The determinant of the covariance matrix is

$$\det(\Gamma) = \sigma_1^2\sigma_2^2(1 - \rho^2),$$

and its inverse is

$$\Gamma^{-1} = \frac{1}{\sigma_1^2\sigma_2^2(1-\rho^2)} \begin{pmatrix} \sigma_2^2 & -\sigma_1\sigma_2\rho \\ -\sigma_1\sigma_2\rho & \sigma_1^2 \end{pmatrix}.$$

Hence, the p.d.f. of a general zero-mean 2D Gaussian random vector is of the form

$$f_{\vec{Y}}(y_1, y_2) = \frac{1}{(2\pi)^{2/2}\sigma_1\sigma_2\sqrt{1-\rho^2}}$$

$$\times \exp\left[-\frac{1}{2}(y_1, y_2) \cdot \frac{\begin{pmatrix} \sigma_2^2 & -\sigma_1\sigma_2\rho \\ -\sigma_1\sigma_2\rho & \sigma_1^2 \end{pmatrix}}{\sigma_1^2\sigma_2^2(1-\rho^2)} \begin{pmatrix} y_1 \\ y_2 \end{pmatrix}\right],$$

which, after performing the prescribed matrix algebra, leads to the final expression:

$$f_{\vec{Y}}(y_1, y_2) = \frac{1}{2\pi\sigma_1\sigma_2\sqrt{1-\rho^2}} \cdot \exp\left[-\frac{1}{2(1-\rho^2)}\left(\frac{y_1^2}{\sigma_1^2} - 2\rho\frac{y_1 y_2}{\sigma_1\sigma_2} + \frac{y_2^2}{\sigma_2^2}\right)\right].$$

$$(8.2.9)$$

The plots of the above densities are bell-shaped surfaces; we saw one example of such a surface in Chap. 3 (Fig. 3.3.1). The level curves of these densities, described by the equations

$$\frac{y_1^2}{\sigma_1^2} - 2\rho\frac{y_1 y_2}{\sigma_1\sigma_2} + \frac{y_2^2}{\sigma_2^2} = \text{const},$$

are ellipses in the (y_1, y_2)-plane, with semiaxes and orientations depending on the parameters ρ, σ_1, and σ_2, representing, respectively, the correlation coefficient between the two components of the Gaussian random vector $\vec{Y}$, and the variances of the first and second components. Figure 8.2.1 shows the level curves of 2D Gaussian densities for four selections of the three above parameters.

8.3 Gaussian Stationary Signals

By definition, a nondegenerate zero-mean random signal $X(t)$ is Gaussian if, for any positive integer N and any selection of sampling times $t_1 < t_2 < \cdots < t_N$, the random vector

$$\vec{X}_{(t_1,\dots,t_N)} = \begin{pmatrix} X(t_1) \\ X(t_2) \\ \vdots \\ X(t_N) \end{pmatrix} \qquad (8.3.1)$$

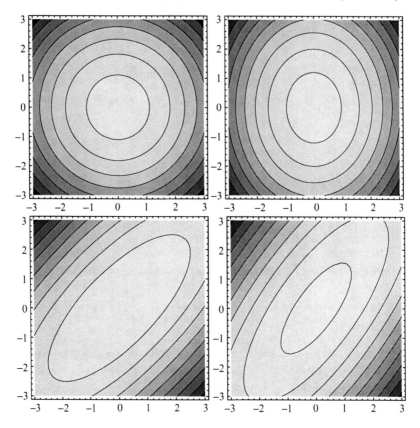

Fig. 8.2.1 Level curves for the 2D Gaussian probability density functions $f_{\vec{Y}}(y_1, y_2)$ (8.2.9), for the following selection of parameters $(\rho, \sigma_1, \sigma_2)$ (clockwise, from the *top left corner*): (0,9,9), (0,8,10), (3/4,9,9), and (3/4,8,10). There are nine level curves in each plot, equally spaced between level zero and the maximum of the p.d.f.

is a Gaussian zero-mean random vector with a nondegenerate covariance matrix. Thus, in view of results of Sect. 8.2, its N-dimensional joint p.d.f. $f_{(t_1,\ldots,t_N)}$ $(x_1, \ldots, x_N)$ is given by the formula[3]

$$f_{(t_1,\ldots,t_N)}(x_1, \ldots, x_N) = \frac{1}{(2\pi)^{N/2} |\det(\mathbf{\Gamma})|^{1/2}} \cdot e^{-\frac{1}{2}\vec{x}^T \mathbf{\Gamma}^{-1} \vec{x}}, \ \det(\mathbf{\Gamma}) \neq 0, \ (8.3.2)$$

where $\mathbf{\Gamma}$ is the $N \times N$ covariance matrix

$$\mathbf{\Gamma} = \mathbf{\Gamma}_{(t_1,\ldots,t_N)} = (\gamma_X(t_i, t_j)) = (\mathbf{E}X(t_i)X(t_j)). \tag{8.3.3}$$

---

[3] Note that for some simple (complex-valued) Gaussian stationary signals, like, e.g., $X(t) = X \cdot e^{jt}$, where $X \sim N(0, 1)$, one can choose the t_i s so that the determinant of the covariance matrix is zero; take, for example, $N = 2$ and $t_1 = \pi, t_2 = 2\pi$. Then the joint p.d.f. of the Gaussian random vector $(X(t_1), \ldots, X(t_N))^T$ is not of the form (8.3.2). Such signals are called degenerate.

Thus, in view of (8.3.1) and (8.3.2), *the only information needed to completely determine all finite-dimensional joint probability distributions of a zero-mean Gaussian random signal $X(t)$ is the knowledge of its autocovariance function,*

$$\gamma_X(s,t) = \mathbf{E}X(s)X(t).$$

For stationary Gaussian signals, the situation is simpler, yet the autocovariance function $\gamma_X(s,t)$ is just a function of a single variable:

$$\gamma_X(s,t) = \gamma_X(t-s).$$

Thus the covariance matrix $\mathbf{\Gamma}$ for a stationary random signal $X(t)$ sampled at $t_1, t_2, \ldots, t_N$ is of the form

$$\mathbf{\Gamma}_{(t_1,\ldots,t_N)} = \begin{pmatrix} \gamma_X(0) & \gamma_X(t_2-t_1) & \gamma_X(t_3-t_1) & \cdots & \gamma_X(t_N-t_1) \\ \gamma_X(t_1-t_2) & \gamma_X(0) & \gamma_X(t_3-t_2) & \cdots & \gamma_X(t_N-t_2) \\ \cdots & \cdots & \cdots & \cdots & \cdots \\ \gamma_X(t_1-t_N) & \gamma_X(t_2-t_N) & \gamma_X(t_3-t_N) & \cdots & \gamma_X(0) \end{pmatrix}.$$

For the real-valued signals under consideration, it is always symmetric because, in that case, the ACvF is an even function, so that $\gamma_X(t_i - t_j) = \gamma_X(t_j - t_i)$. Also, it is obviously invariant under translations, that is, for any t,

$$\mathbf{\Gamma}_{(t_1,\ldots,t_N)} = \mathbf{\Gamma}_{(t_1+t,\ldots,t_N+t)}, \tag{8.3.4}$$

which, in view of (8.3.2) and (8.3.3), implies that all finite-dimensional p.d.f.s of $X(t)$ are also invariant under translations; that is, for any positive integer N, any sampling times $t_1, \ldots, t_N$, and any time shift t,

$$f_{(t_1,\ldots,t_N)}(x_1,\ldots,x_N) = f_{(t_1+t,\ldots,t_N+t)}(x_1,\ldots,x_N). \tag{8.3.5}$$

In other words:

> a Gaussian weakly stationary signal is strictly stationary.

In the particular case when the sampling times are uniformly spaced with the intersampling time interval Δt, the covariance matrix $\mathbf{\Gamma}$ of the signal $X(t)$ sampled at times

$$t, \, t+\Delta t, \, t+2\Delta t, \, \ldots, \, t+(N-1)\Delta t$$

is

$$\begin{pmatrix} \gamma_X(0) & \gamma_X(\Delta t) & \gamma_X(2\Delta t) & \cdots & \gamma_X((N-1)\Delta t) \\ \gamma_X(\Delta t) & \gamma_X(0) & \gamma_X(\Delta t) & \cdots & \gamma_X((N-2)\Delta t) \\ \cdots & \cdots & \cdots & \cdots & \cdots \\ \gamma_X((N-1)\Delta t) & \gamma_X((N-2)\Delta t) & \gamma_X((N-3)\Delta t) & \cdots & \gamma_X(0) \end{pmatrix}.$$

Example 8.3.1 (Ornstein–Uhlenbeck random signal (process)). Consider a Gaussian signal $X(t)$ with autocovariance function

$$\gamma_X(t) = e^{-0.3|t|}.$$

We are interested in finding the joint p.d.f. of the signal at times $t_1 = 1, t_2 = 2$ and the probability that the signal has values between -0.6 and 1.4 at t_1, and between 0.7 and 2.6 at t_2.

The first step is then to find the covariance matrix

$$\Gamma_{(1,2)} = \begin{pmatrix} \gamma_X(0) & \gamma_X(1) \\ \gamma_X(1) & \gamma_X(0) \end{pmatrix} = \begin{pmatrix} e^0 & e^{-0.3} \\ e^{-0.3} & e^0 \end{pmatrix} = \begin{pmatrix} 1 & 0.74 \\ 0.74 & 1 \end{pmatrix}.$$

The covariance coefficient of $X(1)$ and $X(2)$ is then

$$\rho = \frac{\gamma_X(2-1)}{\gamma_X(0)} = 0.74,$$

and, in view of Example 8.2.1 (8.2.9), the joint p.d.f. of $X(1)$ and $X(2)$ is of the form

$$\begin{aligned}
f_{(1,2)}(x_1, x_2) &= \frac{1}{2\pi\sqrt{1-0.74^2}} \cdot \exp\left[\frac{-1}{2(1-0.74^2)}\left(x_1^2 - 2\cdot 0.74 x_1 x_2 + x_2^2\right)\right] \\
&= 0.24 \cdot \exp\left[-1.11\left(x_1^2 - 1.48 x_1 x_2 + x_2^2\right)\right].
\end{aligned}$$

Finally, the desired probability is

$$\mathbf{P}\left(-0.6 \le X(1) \le 1.4 \quad \text{and} \quad 0.7 \le X(2) \le 2.6\right)$$

$$= \int_{-0.6}^{1.4} \int_{0.7}^{2.6} 0.24 \cdot e^{-1.11(x_1^2 - 1.48 x_1 x_2 + x_2^2)} \, dx_1 \, dx_2 = 0.17,$$

where the last integral was evaluated numerically in *Mathematica* with two-digit precision.

8.4 Sample Path Properties of General and Gaussian Stationary Signals

Mean-square continuity and differentiability. It is clear that the local properties of the autocovariance function $\gamma_X(\tau)$ of a stationary signal $X(t)$ affect properties of the sample paths of the signal itself in the mean-square sense, that is, in terms of the behavior of the expectation of the square of the signal's increments, i.e., the variances of the increments.[4] Indeed, with no distributional assumptions on $X(t)$, we have

$$\sigma^2(\tau) = \mathbf{E}(X(t + \tau) - X(t))^2 = 2(\gamma_X(0) - \gamma_X(\tau));$$

the variance of the increment is independent of t. Hence, we have the following result:

A stationary signal $X(t)$ is continuous in the mean-square sense, that is, for any $t > 0$,

$$\lim_{\tau \to 0} \mathbf{E}(X(t + \tau) - X(t))^2 = 0$$

if, and only if, the autocovariance function $\gamma_X(\tau)$ is continuous at $\tau = 0$; that is,

$$\lim_{\tau \to 0} \gamma_X(\tau) = \gamma_X(0).$$

In particular, signals with autocovariance functions $\gamma_X(\tau) = e^{|\tau|}$ or $\gamma_X(\tau) = 1/(1 + \tau^2)$ are mean-square continuous.

A similar mean-square analysis of the limit at $\tau = 0$ of the differential ratio

$$\mathbf{E}\left(\frac{(X(t + \tau) - X(t))}{\tau}\right)^2 = 2\frac{\gamma_X(0) - \gamma_X(\tau)}{\tau^2}$$

shows that a stationary signal with autocovariance function $\gamma_X(\tau) = e^{|\tau|}$ cannot possibly be mean-square differentiable because in this case

$$\lim_{\tau \to 0} \frac{\gamma_X(0) - \gamma_X(\tau)}{\tau^2} = \lim_{\tau \to 0} \frac{1 - e^{-|\tau|}}{\tau^2} = \infty,$$

whereas the differentiability cannot be excluded for the signal with autocovariance $\gamma_X(\tau) = 1/(1 + \tau^2)$ because, in this case,

$$\lim_{\tau \to 0} \frac{\gamma_X(0) - \gamma_X(\tau)}{\tau^2} = \lim_{\tau \to 0} \frac{1 - 1/(1 + \tau^2)}{\tau^2} = 1.$$

Of course, the above brief discussion just verifies the boundedness of the variance of the signal's differential ratio as $\tau \to 0$, not whether the latter has a limit. So let us take a closer look at the issue of the mean-square differentiability of a stationary signal, that is, the existence of the random quantity $X'(t)$, for a fixed t. First, observe that this existence is equivalent to the statement that[5]

---

[4] Recall that the sequence (X_n) of random quantities is said to converge to X, in the mean square, if $\mathbf{E}|X_n - X|^2 \to 0$, as $n \to \infty$.

[5] This argument relies on the so-called Cauchy criterion of convergence for random quantities with finite variance: *A sequence X_n converges in the mean square as $n \to \infty$; that is, there exists a random quantity X such that $\lim_{n\to\infty} \mathbf{E}(X_n - X)^2 = 0$ if and only if $\lim_{n\to\infty} \lim_{m\to\infty}$*

$$\lim_{\tau_1 \to 0} \lim_{\tau_2 \to 0} \mathbf{E} \left(\frac{X(t + \tau_1) - X(t)}{\tau_1} - \frac{X(t + \tau_2) - X(t)}{\tau_2} \right)^2 = 0.$$

But the expression under the limit signs is equal to

$$\mathbf{E} \left(\frac{X(t + \tau_1) - X(t)}{\tau_1} \right)^2 + \mathbf{E} \left(\frac{X(t + \tau_2) - X(t)}{\tau_2} \right)^2$$
$$- 2\mathbf{E} \left(\frac{X(t + \tau_1) - X(t)}{\tau_1} \cdot \frac{X(t + \tau_2) - X(t)}{\tau_2} \right).$$

So the existence of the derivative $X'(t)$ in the mean square is equivalent to the fact that the first two terms converge to $\gamma_{X'}(0)$ and the third to $-2\gamma_{X'}(0)$. But the convergence of the last term means the existence of the limit

$$\lim_{\tau_1 \to 0} \lim_{\tau_2 \to 0} \frac{1}{\tau_1 \tau_2} \mathbf{E} \Big((X(t + \tau_1) - X(t)) \cdot (X(t + \tau_2) - X(t)) \Big)$$
$$= \lim_{\tau_1 \to 0} \lim_{\tau_2 \to 0} \frac{1}{\tau_1 \tau_2} \Big(\gamma_X(\tau_2 - \tau_1) - \gamma_X(\tau_1) - \gamma_X(\tau_2) + \gamma_X(0) \Big)$$
$$= \lim_{\tau_1 \to 0} \lim_{\tau_2 \to 0} \frac{1}{\tau_1 \tau_2} \Delta_{-\tau_1} \Delta_{\tau_2} \gamma_X(0),$$

where $\Delta_\tau f(t) := f(t + \tau) - f(t)$ is the usual difference operator. Indeed,

$$\Delta_{-\tau_1} \Delta_{\tau_2} \gamma_X(0) = \Delta_{\tau_1} (\gamma_X(\tau_2) - \gamma_X(0))$$
$$= (\gamma_X(\tau_2 - \tau_1) - \gamma_X(-\tau_1)) - (\gamma_X(\tau_2) - \gamma_X(0)).$$

Since the existence of the last limit appearing above means twice differentiability of the autocovariance function of X at $\tau = 0$, we arrive at the following criterion:

A stationary signal $X(t)$ is mean-square differentiable if and only if its autocovariance function $\gamma_X(\tau)$ is twice differentiable at $\tau = 0$. Moreover, in this case, the cross-covariance of the signal $X(t)$ and its derivative $X'(t)$ is

$$\mathbf{E}X(t)X'(s) = \lim_{\tau \to 0} \frac{\gamma_X(t + \tau - s) - \gamma_X(t - s)}{\tau} = \frac{\partial}{\partial t} \gamma_X(t - s), \qquad (8.4.1)$$

and the autocovariance of the derivative signal is

$$\mathbf{E}X'(t)X'(s) = \lim_{\tau \to 0} \frac{1}{\tau} \left(\frac{\partial}{\partial t} \gamma_X(t + \tau - s) - \frac{\partial}{\partial t} \gamma_X(t - s) \right) = \frac{\partial^2}{\partial t \, \partial s} \gamma_X(t - s).$$
$$(8.4.2)$$

$\mathbf{E}(X_n - X_m)^2 = 0$. This criterion permits the verification of the convergence without knowing what the limit is; see, e.g., Theorem 11.4.2 in W. Rudin, *Principles of Mathematical Analysis*, McGraw-Hill, New York, 1976.

In a similar fashion, one can calculate the cross-covariance of higher derivatives of the signal $X(t)$ to obtain that[6]

$$\mathbf{E}X^{(n)}(t)X^{(m)}(s) = \frac{\partial^{n+m}}{\partial t^n \partial s^m}\gamma_X(t-s). \tag{8.4.3}$$

Sample path continuity. A study of properties of the individual sample paths (trajectories, realizations) of stationary random signals is a more delicate matter, with the most precise results obtainable only in the case of Gaussian signals. Indeed, we have observed in the previous sections that for a Gaussian signal, the autocovariance function determines all the finite-dimensional probability distributions of the signal, meaning that for any finite sequence of windows $[a_1, b_1], [a_2, b_2], \ldots, [a_N, b_N]$ and any collections of time instants $t_1, t_2, \ldots, t_N$ we can find the probability that the signal fits into those windows at prescribed times; that is,

$$\mathbf{P}(a_1 < t_1 < b_1, \ a_2 < t_2 < b_2, \ldots, \ a_N < t_N < b_N).$$

So it seems that by taking N to ∞, and making the time instants closer to each other, and the windows narrower, one could find the probability that the signal's sample path has any specific shape or property. This idea is, roughly speaking, correct but only in a subtle sense that will be explained below.

The discussion of the sample path properties of stationary signals will be based here on the following theorem of the theory of general random signals (stochastic processes) due to N. N. Kolmogorov:

Theorem 8.4.1. *Let $g(h)$ be an even function, nondecreasing for $h > 0$, and such that $g(h) \to 0$ as $h \to 0$. Furthermore, suppose that $X(t)$ is a random signal such that*

$$\mathbf{P}\Big(|X(t+h) - X(t)| > g(h)\Big) \leq q(h), \tag{8.4.4}$$

for a function $q(h)$ satisfying the following three conditions:

$$q(h) \to 0, \quad \text{as} \quad h \to 0; \tag{8.4.5}$$

$$\sum_{n=1}^{\infty} 2^n q(2^{-n}) < \infty; \tag{8.4.6}$$

$$\sum_{n=1}^{\infty} g(2^{-n}) < \infty. \tag{8.4.7}$$

Then, with probability 1, the sample paths of the signal $X(t)$ are continuous.

---

[6] For details, see M. Loève, *Probability Theory*, Van Nostrand, Princeton, NJ, 1963, Sect. 34.3.

Although the proof of the above theorem is beyond the scope of this book,[7] the intuitive meaning of the assumptions (8.4.4)–(8.4.7) is clear: For the signal to have continuous sample paths, the increments of the signal over small time intervals can be permitted to be large only with a very small probability.

Applied to the second-order (not necessarily stationary) signals, Theorem 8.4.1 immediately gives the following.

Corollary 8.4.1. *If there exists a τ_0 such that, for all τ, $0 \leq \tau < \tau_0$, and all t in a finite time interval,*

$$\mathbf{E}\Big(X(t+\tau) - X(t)\Big)^2 \leq C|\tau|^{1+\epsilon} \tag{8.4.8}$$

for some constants $C, \epsilon > 0$, then the sample paths of the signal $X(t)$ are continuous with probability 1.

To see how Corollary 8.4.1 follows from Theorem 8.4.1, observe first that for any random quantity Z and any constant $a > 0$,[8]

$$\mathbf{P}(Z > a) \leq \int_a^\infty f_Z(z)\,dz \leq \int_a^\infty \frac{z^2}{a^2} f_Z(z)\,dz \leq \frac{\mathbf{E}Z^2}{a^2}.$$

Condition (8.4.8) implies then that

$$\mathbf{P}\Big(X(t+\tau) - X(t)| > g(\tau)\Big) \leq \frac{C|\tau|^{1+\epsilon}}{g^2(\tau)},$$

so that selecting $g(\tau) = |\tau|^{\epsilon/4}$ and

$$q(\tau) = \frac{C|\tau|^{1+\epsilon}}{g^2(\tau)} = C|\tau|^{1+\epsilon/2},$$

we easily see that $g(\tau)$ and $q(\tau)$ are continuous functions vanishing at $\tau = 0$, and that conditions (8.4.4)–(8.4.7) of the theorem are also satisfied. Indeed,

$$\sum_{n=1}^\infty 2^n q(2^{-n}) = C \sum_{n=1}^\infty 2^n (2^{-n})^{1+\epsilon/2} = C \sum_{n=1}^\infty 2^{-n\epsilon/2} < \infty$$

and

$$\sum_{n=1}^\infty g(2^{-n}) = \sum_{n=1}^\infty 2^{-n\epsilon/4} < \infty.$$

---

[7] For a more complete discussion of this theorem and its consequences for the sample path, continuity and differentiability of random signals, see, for example, M. Loève, *Probability Theory*, Van Nostrand, Princeton, NJ, 1963, Sect. 35.3.

[8] This inequality is known as the Chebyshev inequality and its proof here has been carried out only in the case of absolutely continuous probability distributions. The proof in the discrete case is left to the reader as an exercise; see Sect. 8.5.

In the special case of a stationary signal, we have $\mathbf{E}(X(t+\tau)-X(t))^2 = 2(\gamma_X(0) - \gamma_X(\tau))$, so the sample path continuity is guaranteed by the following condition on the autocovariance function:

$$|\gamma_X(0) - \gamma_X(\tau)| \leq C|\tau|^{1+\epsilon}, \tag{8.4.9}$$

for some constant $\epsilon > 0$ and small enough τ.

In particular, for the autocovariance function $\gamma_X(\tau) = 1/(1+\tau^2)$,

$$|\gamma_X(0) - \gamma_X(\tau)| = 1 - \frac{1}{1+\tau^2} = \frac{\tau^2}{1+\tau^2} \leq \tau^2,$$

and condition (8.4.8) is satisfied, thus giving the sample path continuity.

However, for a signal with autocovariance function $\gamma_X(\tau) = e^{-|\tau|}$, the difference $\gamma_X(0) - \gamma_X(\tau)$ behaves asymptotically like τ, for $\tau \to 0$. Therefore, there is no positive ϵ for which condition (8.4.9) is satisfied and we cannot claim the continuity of the sample path in this case – not a surprising result if one remembers that the exponential autocovariance was first encountered in the context of the obviously sample path discontinuous switching signal. Nevertheless, as we observed at the beginning of this section, a signal with an exponential autocovariance is mean-square continuous.

For a Gaussian stationary signal $X(t)$, Theorem 8.4.1 can be applied in a more precise fashion since the probabilities $\mathbf{P}(X(t+\tau) - X(t) > a)$ are known exactly. Indeed, since for any positive z,

$$\int_z^\infty e^{-x^2/2}\,dx \leq \int_z^\infty \frac{x}{z}e^{-x^2/2}\,dx = \frac{1}{z}e^{-z^2/2},$$

because $x/z \geq 1$ in the interval of integration, we have, for any nonnegative function $g(\tau)$ and positive constant C,

$$\mathbf{P}\left(|X(t+\tau) - X(t)| > Cg(\tau)\right) \leq \sqrt{\frac{2}{\pi}}\frac{\sigma(\tau)}{Cg(\tau)}\exp\left(-\frac{1}{2}\frac{C^2g^2(\tau)}{\sigma^2(\tau)}\right), \tag{8.4.10}$$

where $\sigma^2(\tau) = \mathbf{E}(X(t+\tau) - X(t))^2 = 2(\gamma_X(0) - \gamma_X(\tau))$. This estimate yields the following result:

Corollary 8.4.2. *If there exists τ_0 such that for all τ, $0 \leq \tau \leq \tau_0$, the autocovariance function $\gamma_X(\tau)$ of a stationary Gaussian signal $X(t)$ satisfies the condition*

$$\gamma_X(0) - \gamma_X(\tau) \leq \frac{K}{|\ln|\tau||^\delta}, \tag{8.4.11}$$

for some constants $K > 0$ and $\delta > 3$, then the signal $X(t)$ has continuous sample paths with probability 1.

The proof of the corollary is completed by selecting

$$g(\tau) = |\ln|\tau||^{-\nu},$$

with any number ν satisfying condition $1 < \nu < (\delta - 1)/2$, choosing

$$q(C, \tau) = \frac{K'}{C|\ln|\tau||^{\delta/2-\nu}} \exp\left(-\frac{C^2}{2K}|\ln|\tau||^{\delta-2\nu}\right)$$

and verifying the convergence of the two series in conditions (8.4.6)–(8.4.7); see an exercise in Sect. 8.5.

Returning to the case of a stationary random signal with an exponential autocovariance function, we see that if the signal is Gaussian, then Corollary 8.4.2 guarantees the continuity of its sample paths with probability 1. Indeed, condition (8.4.11) is obviously satisfied since (e.g., picking $\delta = 4$) we have

$$\lim_{\tau \to 0}(\gamma_X(0) - \gamma_X(\tau)) \cdot |\ln|\tau||^4 = \lim_{\tau \to 0}(1 - e^{-|\tau|}) \cdot |\ln|\tau||^4 = 0$$

in view of de l'Hôpital's rule.

8.5 Problems and Exercises

8.5.1. A zero-mean Gaussian random signal has the autocovariance function of the form

$$\gamma_X(\tau) = e^{-0.1|\tau|} \cos 2\pi\tau.$$

Plot it. Find the power spectrum $S_X(f)$. Write the covariance matrix for the signal sampled at four time instants separated by 0.5 s. Find its inverse (numerically; use any of the familiar computing platforms, such as *Mathematica, Matlab*, etc.).

8.5.2. Find the joint p.d.f. of the signal from Problem 8.5.3 at $t_1 = 1$ and $t_2 = 2$. Write the integral formula for

$$P(0 \leq X(1) \leq 1, 0 \leq X(2) \leq 2).$$

Evaluate the above probability numerically.

8.5.3. Find the joint p.d.f. of the signal from Problem 8.5.1 at $t_1 = 1, t_2 = 1.5$, $t_3 = 2$, and $t_4 = 2.5$. Write the integral formula for

$$P(-2 \leq X(1) \leq 2, -1 \leq X(1.5) \leq 4, -1 \leq X(2) \leq 1, 0 \leq X(2.5) \leq 3).$$

Evaluate the above probability numerically.

8.5.4. Show that if a 2D Gaussian random vector $\vec{Y} = (Y_1, Y_2)$ has uncorrelated components Y_1, Y_2, then those components are statistically independent random quantities.

8.5.5. Produce 3D surface plots for p.d.f.s of three 2D Gaussian random vectors: $(X(1.0), X(1.1))^T$, $(X(1.0), X(2.0))^T$, $(X(1.0), X(5.0))^T$, where $X(t)$ is the stationary signal described in Example 8.3.1. Comment on similarities and differences in the three plots.

8.5.6. Prove that if there exists a τ_0 such that for all $\tau < \tau_0$ and all t in a finite time interval,

$$\mathbf{E}\Big(X(t+\tau) - X(t)\Big)^2 \leq C \frac{|\tau|}{|\ln|\tau||^{1+\delta}},$$

for some $C > 0$ and $\delta > 2$, then the sample paths of the signal $X(t)$ are continuous with probability 1. *Hint:* This result is a little more delicate than Corollary 8.4.1, but the idea of the proof is similar: Take $g(\tau) = |\ln|\tau||^{-\beta}$, for a β between 1 and $\delta/2$, whereby $q(\tau) = |\tau|/|\ln|\tau||^{1+\delta-2\beta}$, and check conditions (8.4.4)–(8.4.7) in Theorem 8.4.1.

8.5.7. Verify the Chebyshev inequality $\mathbf{P}(|Z| > a) \leq \mathbf{E}Z^2/a^2$, $a > 0$, for a discrete random quantity Z.

8.5.8. Produce 3D plots of several 2D Gaussian densities with selected means and covariance matrices. Then plot level curves for them.

8.5.9. The random signal $X(t)$ has an autocovariance function of the form $\gamma_X(\tau) = \exp(-|\tau|^{\alpha})$ with $0 < \alpha \leq 2$. For which values of parameter α can you claim the continuity of sample paths of $X(t)$ with probability 1? For $\alpha > 2$, the above formula does not give a covariance function of any stationary signal. Why? *Hint:* Check the positive-definiteness condition from Remark 5.2.1.

8.5.10. Verify formula (8.4.3) for the cross-covariance of higher derivatives of a stationary signal.

8.5.11. Verify the convergence of the series (8.4.6)–(8.4.7) in the proof of Corollary 8.4.2.

Chapter 9
Spectral Representation of Discrete-Time Stationary Signals and Their Computer Simulations

Given an arbitrary power spectrum $S_X(f)$ or, equivalently, its inverse Fourier transform, the autocovariance function $\gamma_X(\tau)$, our ability to simulate the corresponding stationary random signals $X(t)$, using only the pseudo-random number generator, which produces, say, discrete-time white noise, depends on the observation that, in some sense, all stationary random signals can be approximated by superpositions of random harmonic oscillations such as those discussed in Examples 4.1.2 and 4.1.9. Recall that if $A_1, \ldots, A_N$ are independent, zero-mean random variables with finite variance, and $f_1, \ldots, f_N$ is a sequence of distinct frequencies, then a random superposition of N simple complex-valued harmonic oscillations in discrete time, $n = \ldots, -1, 0, 1, \ldots,$

$$X_N(n) = \sum_{k=1}^{N} A_k \cdot e^{j2\pi f_k n}, \tag{9.0.1}$$

is a stationary signal with the autocovariance function of the form

$$\gamma_{X_N}(n) = \sum_{k=1}^{N} \mathbf{E}|A_k|^2 \cdot e^{j2\pi f_k n}. \tag{9.0.2}$$

This suggests the following, intuitive approach to our simulation problem: Given a power spectrum $S_X(f)$, concentrated, say, on the frequency interval $[0, 1]$, mimicking the continuous-time analysis of Sect. 5.2, we can expect the corresponding ACvF to be the "discrete-time inverse Fourier transform," i.e., the Fourier coefficients of $S_X(f)$,

$$\gamma_X(n) = \int_0^1 S(f)e^{j2\pi f n}\, df.$$

The latter integral can now be approximated by its discretized version, so that

$$\gamma_X(n) \approx \sum_{k=1}^{N} S(f)\, \Delta f_k\, e^{j2\pi f_k n}, \tag{9.0.3}$$

W.A. Woyczyński, *A First Course in Statistics for Signal Analysis*,
DOI 10.1007/978-0-8176-8101-2_9, © Springer Science+Business Media, LLC 2011

where
$$0 = f_0 < f_1 < \cdots < f_N = 1$$

is a partition of the $[0, 1]$ interval and $\Delta f_k = f_k - f_{k-1}$. Comparing (9.0.2) and (9.0.3), it seems that to produce an approximated version of $X(n)$, it now suffices to generate a standard white noise $W(k), k = 1, \ldots, N$, take as the random amplitudes in (9.0.1) the sequence

$$A_k = \sqrt{S(f_k)\Delta f_k}\, W(k), \qquad k = 1, \ldots, N, \tag{9.0.4}$$

so that $\mathbf{E}|A_k|^2 = S(f_k)\Delta f_k$, and produce the sequence

$$X_N(n) = \sum_{k=1}^{N} \sqrt{S(f_k)\Delta f_k}\, W(k) \cdot e^{j2\pi f_k n} \approx X(n). \tag{9.0.5}$$

Alternatively, we can consider the Fourier series expansion of the power spectrum (see Chap. 2, but here the variable is the frequency f)

$$S_X(f) = \sum_{n=-\infty}^{\infty} c_n \cdot e^{j2\pi f n}, \tag{9.0.6}$$

with the Fourier coefficients

$$c_n = \int_0^1 S_X(f) \cdot e^{-j2\pi f n}\, df. \tag{9.0.7}$$

Now, the above integral can be replaced, approximately, by the discretized sum

$$c_n \approx \sum_{k=1}^{K} a_k \cdot e^{-j2\pi f_k n}, \tag{9.0.8}$$

with the Fourier coefficients

$$a_k = \int_{f_{k-1}}^{f_k} S_X(f)\, df, \qquad k = 1, 2, \ldots, K, \tag{9.0.9}$$

corresponding to the power of the signal $X(n)$ concentrated in each of the frequency bands $[f_{k-1}, f_k], k = 1, \ldots, K$. Finally, we recognize in (9.0.8) the discrete-time version of the ACvF of the form (9.0.2) of the signal of the form (9.0.1), which gives us yet another approximate expression for the sought-after signal $X(n)$:

$$X_K(n) \approx \sum_{k=1}^{K} A_k \cdot e^{-j2\pi f_k n}, \tag{9.0.10}$$

where the A_ks are selected to be arbitrary zero-mean, independent random variables, with $\mathbf{E}|A_k|^2 = a_k, k = 1, \ldots, K$, so that

$$\gamma_{X_K}(n) = \sum_{k=1}^{K} a_k \cdot e^{-j2\pi f_k n} \approx c_n. \qquad (9.0.11)$$

If $W(k)$ is the standard white noise (of an arbitrary distribution), then choosing

$$A_k = \sqrt{a_k} \cdot W(k), \qquad k = 1, \ldots, K, \qquad (9.0.12)$$

will also do the job.

Obviously, the key to applying the above schemes is in the details: In what sense is the approximation meant? What are the precise algorithms? What is the rigorous justification for them? Also, clearly, for smooth spectra, $S(f)$ and large K and N, the difference between expressions (9.0.5) and (9.0.10) is negligible.

In this chapter we work with discrete-time signals, and the rigorous answer to the above questions is contained in the so-called spectral representation theorem for stationary random signals which is derived in this chapter. On the way to its formulation we introduce the necessary concepts, including the crucial construction of *stochastic integrals* with respect to a white noise signal, often called the *white noise integrals*. We conclude with a computer algorithm based on the *spectral representation theorem*.

9.1 Autocovariance as a Positive-Definite Sequence

In this chapter we will study random stationary signals in discrete time, that is, sequences of *complex-valued* random quantities

$$\ldots, X(-2), X(-1), X(0), X(1), X(2), \ldots,$$

with time n extending all the way from minus to plus infinity. The stationarity is meant in the second-order weak sense; that is, we will assume that the means $\mathbf{E}X(n) = 0$ and the autocovariance function, now really a sequence,

$$\mathbf{E}[X^*(m)X(n)] = \gamma(n-m), \qquad m, n = \ldots, -2, -1, 0, 1, 2, \ldots,$$

depends only on the time lag $\tau = n - m$. The following properties of the *autocovariance sequence* are immediately verified:

For any n,

$$\mathbf{E}|X(n)|^2 = \mathbf{E}[X^*(n)X(n)] = \mathbf{E}|X(0)|^2 = \gamma_X(0) \geq 0, \qquad (9.1.1)$$

$$\gamma_X(-n) = \gamma_X^*(n), \qquad (9.1.2)$$

$$|\gamma_X(n)| \leq \gamma_X(0). \qquad (9.1.3)$$

The last inequality is a direct consequence of the Cauchy–Schwartz inequality.

Also, importantly, the autocovariance sequence is *positive-definite*; that is, for any positive integer N, arbitrary integers, $n_1, n_2, \ldots, n_N$, and arbitrary complex numbers $\lambda_1, \lambda_2, \ldots, \lambda_N$,

$$\sum_{i,k=1}^{N} \gamma_X(n_i - n_k)\lambda_i \lambda_k^* \geq 0. \tag{9.1.4}$$

Indeed,

$$\sum_{i,k=1}^{N} \gamma_X(n_i - n_k)\lambda_i \lambda_k^*$$

$$= \sum_{i,k=1}^{N} \mathbf{E}[X(n_i)X^*(n_k)]\lambda_i \lambda_k^* = \mathbf{E} \sum_{i,k=1}^{N} [\lambda_i X(n_i)] \cdot [\lambda_k X(n_k)]^*$$

$$= \mathbf{E} \sum_{i=1}^{N} \lambda_i X(n_i) \cdot \sum_{k=1}^{N} [\lambda_k X(n_k)]^* = \mathbf{E} \left| \sum_{i=1}^{N} \lambda_i X(n_i) \right|^2 \geq 0.$$

Recall (see Remark 5.2.1) that the ACvF in continuous time was also proven to be positive-definite.

9.2 Cumulative Power Spectrum of Discrete-Time Stationary Signal

The development of this section will be analogous to the development of the concept of the power spectrum of continuous-time signals in Sect. 5.2. However, we will proceed in a slightly different fashion, and with more mathematical precision. The basic structural result regarding the autocovariance function of a discrete-time stationary signal can be formulated as follows:

Herglotz's theorem. The following statements about the sequence $\gamma(n)$, $n = \ldots, -2, -1, 0, 1, 2, \ldots$, of complex numbers are equivalent:

(i) The sequence $\gamma(n)$ is an autocovariance sequence of a stationary discrete-time signal; that is, there exists a stationary signal $X(n)$ such that $\gamma(n) = \gamma_X(n)$.
(ii) The sequence $\gamma(n)$ is positive-definite; that is, it satisfies condition (9.0.4).
(iii) There exists a nondecreasing bounded function $\mathcal{S}_X(f)$, defined on the interval $[0, 1]$, such that

$$\gamma(n) = \int_0^1 e^{j2\pi nf} \, d\mathcal{S}(f), \qquad n = \ldots, -2, -1, 0, 1, 2, \ldots. \tag{9.2.1}$$

The function $\mathcal{S}_X(f)$ is called the cumulative power spectrum of the signal X.

Remark 9.2.1 (Power spectrum density). The integral of the form $\int a(f) \, dS(f)$, called the *Stieltjes integral*, is to be understood as the limit of sums $\sum a(f_i) \cdot \Delta S(f_i)$ when $\max_i |\Delta S(f_i)| = S(f_i) - S(f_{i-1}) \to 0$. As before, $0 = f_0 < f_1 < \cdots < f_N = 1$ stands for a partition of the interval $[0, 1]$.

If the cumulative power spectrum has a spectral density $S(f), 0 \leq f \leq 1$, that is,

$$S(f) = \int_0^f S(g) \, dg, \qquad \frac{dS(f)}{df} = S(f) \geq 0,$$

then formula (9.1.1) takes the form of the usual Riemann integral

$$\gamma(n) = \int_0^1 e^{j2\pi nf} S(f) \, df, \qquad n = \ldots, -2, -1, 0, 1, 2, \ldots, \qquad (9.2.2)$$

and the sequence $\gamma(-n)$ can be viewed simply as the sequence of Fourier coefficients of the power spectrum density $S(f)$.

In the special case when the cumulative power spectrum is constant, except for jumps, that is,

$$S(f) = \sum_k s_k \, u(f - f_k), \qquad 0 = f_0 < f_1 < \cdots < f_N = 1,$$

where $u(t)$ is the unit step function, then

$$\int a(f) \, dS(f) = \sum_l a(f_k) s_k,$$

so that

$$\gamma(n) = \sum_k s_k e^{j2\pi nf_k}, \qquad n = \ldots, -2, -1, 0, 1, 2, \ldots, \qquad (9.2.3)$$

and the power spectrum density can be understood as a sum of the Dirac deltas:

$$S(f) = \sum_k s_k \, \delta(f - f_k).$$

However, it is worth remembering that there are so-called singular cumulative power spectra that are not of either of the two types described above (nor their mixtures).[1]

---

[1] See Sect. 3.1 or, e.g., M. Denker and W.A. Woyczyński, *Introductory Statistics and Random Phenomena. Uncertainty, Complexity and Chaotic Behavior in Engineering and Science*, Birkhäuser, Boston, Cambridge, MA, 1998.

Proof of Herglotz's theorem. The implication *(i)* $\Longrightarrow$ *(ii)* has been proved following the definition (9.1.4).

We shall now prove that *(ii)* $\Longrightarrow$ *(iii)*. Assume that $\gamma(n)$ is positive-definite. In view of (9.1.4), selecting $n_i = i$, $\lambda_i = e^{-j2\pi if}$, $i = 1, 2, \ldots, N$, we have

$$0 \le \sum_{i,k=1}^{N} \gamma(i-k) e^{-j2\pi if} e^{j2\pi kf} = \sum_{i,k=1}^{N} \gamma(i-k) e^{-j2\pi(i-k)f}$$

$$= \sum_{m=-N+1}^{N-1} (N - |m|) \gamma(m) e^{-j2\pi mf},$$

after the substitution $m = i - k$. Define

$$S_N(f) := \frac{1}{N} \sum_{m=-N+1}^{N-1} (N - |m|) \gamma(m) e^{-j2\pi mf}.$$

Then

$$S_N(f) \ge 0 \qquad \text{and} \qquad \int_0^1 S_N(f)\, df = \gamma(0). \qquad (9.2.4)$$

By a fundamental real analysis result called the Arzelà–Ascoli theorem,[2] conditions (9.2.4) guarantee the existence of a function $S(f)$ and a sequence $N_i \nearrow \infty, i \to \infty$, such that for each bounded and smooth function $a(f)$,

$$\int_0^1 a(f) S_{N_i}(f)\, df \longrightarrow \int_0^1 a(f)\, dS(f).$$

Therefore, selecting $a(f) = e^{j2\pi mf}$, we have

$$\int_0^1 e^{j2\pi mf}\, dS(f) = \lim_{i \to \infty} \int_0^1 e^{j2\pi mf} S_{N_i}(f)\, df = \gamma(m)$$

because, for each m such that $|m| \le N_i$,

$$\int_0^1 e^{j2\pi mf} S_{N_i}(f)\, df = \gamma(m) \left(1 - \frac{|m|}{N_i}\right).$$

Thus, the existence of the cumulative spectral measure for each discrete-time stationary signal has been established.

---

[2] See, e.g., G. B. Folland, *Real Analysis*, Wiley, New York, 1984.

The implication *(iii)* $\Longrightarrow$ *(ii)* can be verified directly. Indeed, given assumption *(iii)*,

$$\sum_{i,k=1}^{N} \gamma(n_i - n_k)\lambda_i \lambda_k^* = \sum_{i,k=1}^{N} \int_0^1 e^{j2\pi(n_i-n_k)f} \, dS(f) \cdot \lambda_i \lambda_k^*$$

$$= \int_0^1 \sum_{i,k=1}^{N} \left[\lambda_i e^{j2\pi n_i f}\right] \cdot \left[\lambda_k e^{j2\pi n_k f}\right]^* dS(f)$$

$$= \int_0^1 \left| \sum_{i=1}^{N} \lambda_i e^{j2\pi n_i f} \right|^2 dS(f) \geq 0,$$

because $S(f)$ is nondecreasing, so that its increments, "$dS(f)$," are nonnegative.

The implication *(ii)* $\Longrightarrow$ *(i)* follows from the following fact established in Sect. 8.2. For any given positive-definite matrix $\Gamma = (\gamma_{ik})$, $i, k = 1, 2, \ldots, N$, there exists a Gaussian random vector $X = (X_1, X_2, \ldots, X_N)$, with covariance matrix Γ. Now, for any N, it suffices to take $\Gamma = (\gamma(i - k))$, $i, k = 1, 2, \ldots, N$, and define

$$X(1) = X_1, \ X(2) = X_2, \ \ldots, \ X(N) = X_N.$$

This proves the existence of a finite discrete-time stationary random signal with an autocovariance sequence given by a prescribed positive-definite sequence.[3]

9.3 Stochastic Integration with Respect to Signals with Uncorrelated Increments

Recall that our goal in this chapter is to develop a simulation algorithm for discrete-time stationary signals with a given power spectrum. One of the methods used for that purpose involves representation of the random signal as a stochastic integral with respect to another random signal which has uncorrelated increments which is easy to simulate via a pseudo-random number generator. The purpose of this section is to introduce such integrals.

The finite variance, zero-mean, real-valued signal $\mathcal{W}(w)$ of continuous or discrete parameter w is said to have *uncorrelated increments* if, for any $w_1 \leq w_2 \leq w_3$,

$$\mathbf{E}[(\mathcal{W}(w_3) - \mathcal{W}(w_2)) \cdot (\mathcal{W}(w_2) - \mathcal{W}(w_1))] = 0. \qquad (9.3.1)$$

In other words, such signals have uncorrelated increments over disjoint intervals of parameter w. Observe that condition (9.3.1) can be rewritten in terms of the auto-covariance function $\gamma_{\mathcal{W}}(v, w) = \mathbf{E}\mathcal{W}(v)\mathcal{W}(w)$ (which here is truly a function of

---

[3] A step proving the existence of an *infinite* such sequence requires an application of the so-called Kolmogorov extension theorem; see, e.g., P. Billingsley, *Probability and Measure*, Wiley, New York, 1986.

two variables v, w, and not just the parameter lag $w - v$, as is the case for stationary signals) as follows:

$$\mathbf{E}[(\mathcal{W}(w_3) - \mathcal{W}(w_2)) \cdot (\mathcal{W}(w_2) - \mathcal{W}(w_1))]$$
$$= \mathbf{E}\mathcal{W}(w_3)\mathcal{W}(w_2) - \mathbf{E}\mathcal{W}(w_2)\mathcal{W}(w_2) - \mathbf{E}\mathcal{W}(w_3)\mathcal{W}(w_1) + \mathbf{E}\mathcal{W}(w_2)\mathcal{W}(w_1)$$
$$= \gamma_\mathcal{W}(w_3, w_2) - \gamma_\mathcal{W}(w_2, w_2) - \gamma_\mathcal{W}(w_3, w_1) + \gamma_\mathcal{W}(w_2, w_1) = 0. \qquad (9.3.2)$$

Example 9.3.1 (Random walk: The cumulative white noise in discrete time). In discrete time, the white noise, $W(n)$, was defined simply as a sequence of zero-mean, independent (and thus uncorrelated), identically distributed random quantities with finite variance, so that its autocovariance sequence is

$$\gamma_W(n, m) = \gamma_W(m - n) = \mathbf{E}W(n)W(m) = \begin{cases} 0, & \text{if } n - m \neq 0; \\ \sigma^2, & \text{if } n - m = 0. \end{cases}$$

We will define the *random walk*, or *cumulative white noise*, generated by the white noise $W(n)$ as the random signal

$$\mathcal{W}(n) = W(1) + W(2) + \cdots + W(n), \qquad n = 1, 2, \ldots,$$

with the convention $\mathcal{W}(0) = 0$.

The following mental picture is worth keeping in mind: In the case of the symmetric Bernoulli white noise $W(n)$, with $\mathbf{P}(W(n) = \pm 1) = 1/2$, the generated random walk $\mathcal{W}(n)$ moves "forward" by 1 whenever $W(n) = +1$, and "backward" by 1 whenever $W(n) = -1$; each possibility occurs with probability 1/2.

The cumulative white noise has uncorrelated increments. Indeed, if $n_1 \leq n_2 \leq n_3$, then

$$\mathbf{E}[(\mathcal{W}(n_3) - \mathcal{W}(n_2)) \cdot (\mathcal{W}(n_2) - \mathcal{W}(n_1))]$$
$$= \mathbf{E}\left[\left(\sum_{n=1}^{n_3} W(n) - \sum_{n=1}^{n_2} W(n)\right) \cdot \left(\sum_{n=1}^{n_2} W(n) - \sum_{n=1}^{n_1} W(n)\right)\right]$$
$$= \mathbf{E}[(W(n_2 + 1) + \cdots + W(n_3)) \cdot (W(n_1 + 1) + \cdots + W(n_2))]$$
$$= \mathbf{E}(W(n_1 + 1) + \cdots + W(n_2)) \cdot \mathbf{E}(W(n_2 + 1) + \cdots + W(n_3)) = 0,$$

because $W(n_1 + 1) + \cdots + W(n_2)$ and $W(n_2 + 1) + \cdots + W(n_3)$ are independent and zero-mean.

For any signal $\mathcal{W}(w)$ with uncorrelated increments, we will introduce a *cumulative control function*

$$C(w) := \mathbf{E}[\mathcal{W}(w) - \mathcal{W}(0)]^2 = \mathbf{E}[\mathcal{W}(w)]^2 \geq 0, \qquad (9.3.3)$$

which simply measures the variance of the increment of the signal from 0 to w. Since the variance of the sum of uncorrelated random quantities is the sum of their variances, the cumulative control function is always nondecreasing because, for $0 \le v \le w$,

$$C(w) = \mathbf{E}[(\mathcal{W}(w) - \mathcal{W}(0)]^2 = \mathbf{E}[(\mathcal{W}(w) - \mathcal{W}(v)) + (\mathcal{W}(v) - W(0))]^2$$
$$= \mathbf{E}[(\mathcal{W}(w) - \mathcal{W}(v))]^2 + \mathbf{E}[(\mathcal{W}(v) - W(0))]^2 \ge \mathbf{E}[(\mathcal{W}(v) - W(0))]^2 = C(v).$$
$$(9.3.4)$$

Observe that, under condition $\mathcal{W}(0) = 0$, the cumulative control function determines the correlation structure of $\mathcal{W}(w)$, and vice versa. If, say, $0 \le v \le w$, then

$$\gamma_{\mathcal{W}}(v, w) = \mathbf{E}\mathcal{W}(v)\mathcal{W}(w)$$
$$= \mathbf{E}[\mathcal{W}(v) - \mathcal{W}(0)] \cdot [(\mathcal{W}(w) - \mathcal{W}(v)) + (\mathcal{W}(v) - \mathcal{W}(0))]$$
$$= \mathbf{E}[\mathcal{W}(v) - \mathcal{W}(0)] \cdot [(\mathcal{W}(v) - \mathcal{W}(0))] = C(v),$$

because the increments over the intervals $[0, v]$ and $[v, w]$ are uncorrelated. Since an analogous reasoning holds true in the case $0 \le w \le v$, we get the general formula

$$\gamma_{\mathcal{W}}(v, w) = C(\min(v, w)). \qquad (9.3.5)$$

An important class of signals with independent (and thus uncorrelated) increments are those that also have *stationary increments*, that is, for which the c.d.f. of the increment $\mathcal{W}(w) - \mathcal{W}(v)$ is the same as the c.d.f. of the increment $\mathcal{W}(w + z) - \mathcal{W}(v + z)$, for any z. The random walk from Example 9.3.1 is such a signal. For signals with independent and stationary increments, the cumulative control function satisfies condition

$$C(w + v) = C(w) + C(v) \qquad (9.3.6)$$

because

$$\mathbf{E}[\mathcal{W}(w + v) - \mathcal{W}(0)]^2 = \mathbf{E}[\mathcal{W}(w + v) - \mathcal{W}(v)]^2 + \mathbf{E}[\mathcal{W}(v) - \mathcal{W}(0)]^2$$
$$= \mathbf{E}[\mathcal{W}(w) - \mathcal{W}(0)]^2 + \mathbf{E}[\mathcal{W}(v) - \mathcal{W}(0)]^2.$$

Condition (9.3.6) forces the cumulative function to be linear, that is, of the form

$$C_{\mathcal{W}}(w) = \text{const} \cdot w, \qquad (9.3.7)$$

and, in view of (9.3.5), the autocovariance structure of a signal with stationary and uncorrelated increments is of the form

$$\gamma_W(v, w) = \text{const} \cdot \min(v, w). \qquad (9.3.8)$$

Example 9.3.2 (The Wiener, or Brownian motion process). A continuous-time Gaussian signal with stationary and independent increments with

$$C_W(w) = w, \qquad \gamma_W(v, w) = \min(v, w),$$

is called the Wiener stochastic process (or the Brownian motion process). Its sample trajectories are shown in Fig. 1.1.4. Notice that in this case, in view of Sect. 8.3, the autocovariance function gives a complete description of all finite-dimensional distributions of $W(w)$. Indeed, given parameter values

$$w_1 \leq w_2 \leq \cdots \leq w_N,$$

the random vector

$$(W(w_1), W(w_2), \ldots, W(w_N))$$

is a Gaussian random vector with the covariance matrix $\Gamma = (\min(w_i, w_k))$, so that its joint c.d.f. can be explicitly calculated:

$$\mathbf{P}\Big(W(w_1) \leq a_1, W(w_2) \leq a_2, \ldots, W(w_N) \leq a_N\Big)$$

$$= \int_{-\infty}^{a_1} \int_{-\infty}^{a_2} \cdots \int_{-\infty}^{a_N} \frac{e^{-\frac{\zeta_1^2}{2w_1}}}{\sqrt{2\pi w_1}} \cdot \frac{e^{-\frac{(\zeta_2 - \zeta_1)^2}{2(w_2 - w_1)}}}{\sqrt{2\pi(w_2 - w_1)}} \cdots \frac{e^{-\frac{(\zeta_N - \zeta_{N-1})^2}{2(w_N - w_{N-1})}}}{\sqrt{2\pi(w_N - w_{N-1})}}$$

$$\times \, d\zeta_N \cdots d\zeta_2 \cdot d\zeta_1. \tag{9.3.9}$$

At this point, we are able to introduce the stochastic integral

$$\int_0^1 x(w) \, dW(w),$$

with respect to a signal $W(w)$ with uncorrelated increments, for a deterministic, possibly complex-valued, function $x(w)$. If $x(w)$ is a step function of the form

$$x(w) = \sum_{i=1}^N x_i \mathbf{1}_{(w_{i-1}, w_i]}(w), \tag{9.3.10}$$

with $0 = w_0 < w_1 < \cdots < w_{N-1} < w_N = 1$, and $\mathbf{1}_A(w)$ denoting the indicator function of set A,[4] then, obviously,

$$\int_0^1 x(w) \, dW(w) := \sum_{i=1}^N x_i \cdot (W(w_i) - W(w_{i-1})). \tag{9.3.11}$$

---

[4] Recall that the indicator function $\mathbf{1}_A(w)$ is defined as being equal to 1 for w belonging to set A, and being 0 for w outside A.

Note that the variance of the stochastic integral in (9.3.11) is

$$
\mathbf{E}\left|\int x(w)\,d\mathcal{W}(w)\right|^2 = \mathbf{E}\left|\sum_{i=1}^{N} x_i \cdot (\mathcal{W}(w_i) - \mathcal{W}(w_{i-1}))\right|^2
$$

$$
= \sum_{i=1}^{N} |x_i|^2 \mathbf{E}(\mathcal{W}(w_i) - \mathcal{W}(w_{i-1}))^2
$$

$$
= \sum_{i=1}^{N} |x_i|^2 (C(w_i) - C(w_{i-1}))
$$

$$
= \int_0^1 |x(w)|^2\,dC(w), \qquad (9.3.12)
$$

because, in view of (9.3.3), for any $0 < v < w$,

$$
\mathbf{E}(\mathcal{W}(w) - \mathcal{W}(v))^2 = C(w) - C(v). \qquad (9.3.13)
$$

Since any function $x(w)$ such that

$$
\int_0^1 |x(w)|^2\,dC(w) < \infty \qquad (9.3.14)
$$

is a limit of a sequence $x_n(w)$ of step functions,[5] in the sense that

$$
\int_0^1 |x_n(w) - x(w)|^2\,dC(w) \to 0, \quad \text{as} \quad n \to \infty,
$$

definition (9.3.11) of the stochastic integral for step functions can now be extended to any $x(w)$ satisfying condition (9.3.14), that is, square integrable with respect to $dC(w)$, by setting

$$
\int_0^1 x(w)\,d\mathcal{W}(w) := \lim_{n\to\infty} \int_0^1 x_n(w)\,d\mathcal{W}(w), \qquad (9.3.15)
$$

where the limit is understood as the limit in the mean square of random quantities (that is, variance, given that all the random quantities have zero means). In view of this procedure, the general stochastic integral for a function $x(w)$ satisfying condition (9.3.14) enjoys the "isometric" property

$$
\mathbf{E}\left|\int_0^1 x(w)\,d\mathcal{W}(w)\right|^2 = \int_0^1 |x(w)|^2\,dC(w). \qquad (9.3.16)
$$

---

[5] See, e.g., G. B. Folland, *Real Analysis*, Wiley, New York, 1984.

Example 9.3.3 (Gaussian stochastic integrals). Note that if the cumulative control function $C(w)$ of a Gaussian process with independent increments $V(w)$ has a density $c(w)$, that is,

$$C(w) = \int_0^w c(v)\, dv, \qquad \frac{dC(w)}{dw} = c(w) \geq 0, \quad 0 \leq w \leq 1,$$

then, in view of (9.3.16),

$$\mathbf{E}(V(w))^2 = \mathbf{E}\left(\int_0^w dV(v)\right)^2 = \int_0^w c(v)\, dv = \int_0^w \left(\sqrt{c(v)}\right)^2 dv,$$

which implies that for any $x(w)$ satisfying (9.3.14), the statistical properties of the stochastic integrals,

$$\int_0^1 x(v)\, dV(v) \qquad \text{and} \qquad \int_0^1 x(w)\sqrt{c(w)}\, dW(w), \tag{9.3.17}$$

where $W(w)$ is the Wiener process, are the same. Later on this fact will serve as the basis of a computer simulation of stationary random signals with a given spectrum.

Because, for any complex numbers ξ, η, we have the so-called polarization formulas,

$$\mathrm{Re}\,[\xi \cdot \eta^*] = \frac{1}{4}(|\xi + \eta|^2 - |\xi - \eta|^2),$$

$$\mathrm{Im}\,[\xi \cdot \eta^*] = \frac{1}{4}(|\xi + j\eta|^2 - |\xi - j\eta|^2),$$

which express the product in terms of the squared moduli, the "isometric" relation (9.3.16) extends from the mean squares to scalar products. In other words, for any $x(w), y(w)$ satisfying condition (9.3.14),

$$\mathbf{E}\left[\int_0^1 x(w)\, dW(w) \cdot \left(\int_0^1 y(w)\, dW(w)\right)^*\right] = \int_0^1 x(w) \cdot y^*(w)\, dC(w). \tag{9.3.18}$$

9.4 Spectral Representation of Stationary Signals

The fundamental result about the structure of discrete-time stationary signals is that they are, essentially, sequences of random Fourier coefficients of stochastic processes with uncorrelated increments. More precisely, we have the following

Spectral representation theorem. *A discrete-time random signal $X(n)$, $n = \ldots,$ $-2, -1, 0, 1, 2, \ldots$, is stationary if and only if it has the representation*

$$X(n) = \int_0^1 e^{j2\pi nf} \, dW(f) \tag{9.4.1}$$

for a certain random process $W(f)$, $0 \leq f \leq 1$, which has uncorrelated increments. Moreover, the cumulative spectral function of $X(n)$ is identical to the cumulative control function of $W(f)$; that is,

$$S_X(f) = C_W(f), \qquad 0 \leq f \leq 1. \tag{9.4.2}$$

Proof. If the random signal $X(n)$ is of the form (9.4.1), then it is stationary because it has zero mean and because, in view of the "isometry" (9.3.17),

$$E[X(n)X^*(m)] = \mathbf{E}\left[\int_0^1 e^{j2\pi nf} \, dW(f) \cdot \left(\int_0^1 e^{j2\pi mf} \, dW(f) \right)^* \right]$$

$$= \int_0^1 e^{j2\pi(n-m)f} \, dC_W(f).$$

The above calculation also identifies the *cumulative control function* of the process $W(f)$ as the cumulative spectral function of the random signal $X(n)$.

The proof of the reverse implication is more delicate, as it requires identification, for each signal $X(n)$, of a process $W(f)$ yielding representation (9.4.1). So assume that $X(n)$ is a stationary signal with autocovariance sequence

$$\gamma_X(n) = \int_0^1 e^{j2\pi nf} \, dS_X(f).$$

Denote by $L_0^2(\mathbf{P})$ the space of random quantities with zero mean and finite variance into the space $L^2(dS_X(f))$ of complex functions on $[0, 1]$ which are square integrable with respect to the cumulative spectral function $S_X(f)$. Next, consider a linear mapping I from $L_0^2(\mathbf{P})$ into $L^2(dS_X(f))$ defined by the identity

$$I[X(n)] := e^{j2\pi nf}, \qquad n = \ldots, -2, -1, 0, 1, 2, \ldots, \tag{9.4.3}$$

on complex exponentials and extended, in a natural way, to all their combinations. In other words, for any complex numbers $c_{-N}, \ldots, c_{-1}, c_0, c_1, \ldots, c_N$,

$$I\left[\sum_{n=-N}^{N} c_n X(n) \right] = \sum_{n=-N}^{N} c_n e^{j2\pi nf}. \tag{9.4.4}$$

The mapping I is an *isometry*[6] on such linear combinations because

$$\mathbf{E}\left|\sum_{n=-N}^{N} c_n X(n)\right|^2 = \sum_{n,m=-N}^{N} c_n c_m^* \mathbf{E}[X(n) X^*(m)]$$

$$= \sum_{n,m=-N}^{N} c_n c_m^* \int_0^1 e^{j2\pi(n-m)f} \, d\mathcal{C}_W(f)$$

$$= \int_0^1 \left|\sum_{n=-N}^{N} c_n e^{j2\pi nf}\right|^2 d\mathcal{S}_X(f),$$

and, as such, it extends to the linear isometry

$$I : \mathcal{L}[X(n), n = \ldots, -2, -1, 0, 1, 2, \ldots] \longmapsto L^2(d\mathcal{S}_X(f)),$$

where $\mathcal{L}[X(n), n = \ldots, -2, -1, 0, 1, 2, \ldots]$ is the subspace of $L^2(\mathbf{P})$ consisting of linear combinations of $X(n)$s and their mean-square limits. Since any isometry is necessarily a one-to-one mapping, I has a well-defined inverse:

$$I^{-1} : L^2(d\mathcal{S}_X(f)) \longmapsto \mathcal{L}[X(n), n = \ldots, -2, -1, 0, 1, 2, \ldots],$$

which is also a linear isometry.   □

Now we will define a stochastic process $\mathcal{W}(f)$ by the formula

$$\mathcal{W}(f) := I^{-1}(\mathbf{1}_{[0,f]}),$$

where $\mathbf{1}_{[0,f]}(g), 0 \le g \le 1$, is the indicator function of the interval $[0, f]$. This process has zero mean and uncorrelated increments since, for $f_1 \le f_2 \le f_3$, in view of the isometric property of I^{-1},

$$\mathbf{E}[(\mathcal{W}(f_3) - \mathcal{W}(f_2)) \cdot (\mathcal{W}(f_2) - \mathcal{W}(f_1))]$$

$$= \mathbf{E}[(I^{-1}(\mathbf{1}_{[0,f_3]}) - I^{-1}(\mathbf{1}_{[0,f_2]})) \cdot (I^{-1}(\mathbf{1}_{[0,f_2]}) - I^{-1}(\mathbf{1}_{[0,f_1]}))]$$

$$= \mathbf{E}[(I^{-1}(\mathbf{1}_{[0,f_3]}) - (\mathbf{1}_{[0,f_2]})) \cdot (I^{-1}(\mathbf{1}_{[0,f_2]}) - \mathbf{1}_{[0,f_1]}))]$$

$$= \mathbf{E}[I^{-1}(\mathbf{1}_{(f_2,f_3]}) \cdot I^{-1}(\mathbf{1}_{(f_1,f_2]})]$$

$$= \int_0^1 \mathbf{1}_{(f_2,f_3]}(f) \cdot \mathbf{1}_{(f_1,f_2]}(f) \, d\mathcal{C}_X(f) = 0.$$

---

[6] In the sense that it preserves the norms: The standard deviation is in space $L_0^2(\mathbf{P})$, and $\|a\| = (\int_0^1 |a(f)|^2 \, d\mathcal{S}_X(f))^{1/2}$, for an $a(f)$ in $L^2(d\mathcal{S}_X(f))$.

The same calculation shows that

$$EW^2(f) = \int_0^1 1^2_{[0,f]}(g)\, dC_X(g) = C_X(f).$$

Now, proceeding again via step functions as in Sect. 9.2, using the linearity and isometry properties of I^{-1}, we have, for any function $a(f)$ in space $L^2(dC_X(f))$,

$$I^{-1}(a) = \int_0^1 a(f)\, dW(f).$$

In particular, selecting $a(f) = e^{j2\pi nf}$, we obtain

$$X(n) = I^{-1}(e^{j2\pi nf}) = \int_0^1 e^{j2\pi nf}\, dW(f),$$

which concludes the proof of the spectral representation theorem.

Example 9.4.1 (Spectral representation of white noise). Let $W(f)$ be the Wiener process. Its cumulative control function

$$C_W(f) = f = \int_0^f df$$

has a control density function $C_W(f) \equiv 1$. The stationary, discrete-time signal

$$X(n) = \int_0^1 e^{j2\pi nf}\, dW(f)$$

has the spectral density function $S_X(f) = C_W(f) \equiv 1$, and the autocovariance sequence is

$$\gamma_X(n) = EX(n)X^*(0) = \int_0^1 e^{j2\pi nf}\, df = \delta(n) = \begin{cases} 0, & \text{if } n \neq 0; \\ 1, & \text{if } n = 0. \end{cases}$$

Hence, $X(n)$ is the discrete-time white noise discussed in Chap. 5.

Example 9.4.2 (Spectral representation of filtered white noise). Let $X(n)$ be the white noise discussed above. Consider the (acausal) filtered (i.e., moving average of) white noise

$$Y(n) = \sum_{k=-\infty}^{\infty} c_k X(n-k) = \int_0^1 \left(\sum_{k=-\infty}^{\infty} c_k e^{j2\pi(n-k)f} \right) dW(f),$$

for $n = \ldots, -2, -1, 0, 1, 2, \ldots$. Its autocovariance sequence is

$$
\begin{aligned}
\gamma_Y(n) &= \mathbf{E} Y(n) Y^*(0) \\
&= \mathbf{E} \left(\sum_{k=-\infty}^{\infty} c_k X(n-k) \cdot \sum_{k=-\infty}^{\infty} c_k^* X^*(-k) \right) \\
&= \mathbf{E} \sum_{k,l=-\infty}^{\infty} c_k c_l^* X(n-k) X^*(-l) = \sum_{k,l=-\infty}^{\infty} c_k c_l^* \delta(n-(k-l)) \\
&= \sum_{k,l=-\infty}^{\infty} c_k c_l^* \int_0^1 e^{j2\pi(n-(k-l))f} \, df = \int_0^1 |c(f)|^2 e^{j2\pi nf} \, df,
\end{aligned}
$$

where

$$
c(f) = \sum_{k=-\infty}^{\infty} c_k e^{-j2\pi kf},
$$

is well defined as long as $\sum_{k=-\infty}^{\infty} |c_k|^2 < \infty$. Hence, the power spectral density of the filtered white noise is

$$
S_Y(f) = |c(f)|^2.
$$

9.5 Computer Algorithms: Complex-Valued Case

Given a spectral density $S_X(f)$ of a discrete-time, stationary Gaussian signal $X(n)$, we can simulate a sample path of $X(n), n = 1, 2, \ldots, N$, by first calculating the autocovariance function $\gamma_X(n)$ using formula (9.2.2),

$$
\gamma_X(n) = \int_0^1 e^{j2\pi nf} S_X(f) \, df, \tag{9.5.1}
$$

and then by producing a sample of an N-dimensional Gaussian random vector $\mathbf{X} = (X_1, X_2, \ldots, X_n)$, with the covariance matrix $\Gamma = (\gamma_X(n-m), n, m = 1, 2, \ldots, N)$, using the standard statistical software. This, however, would be computationally expensive, and even infeasible if n is large.

So in this section we will describe a different, explicit algorithm for such a simulation based on the spectral representation of Sect. 9.4. The algorithm is mathematically justified by the discussions of the preceding sections, and it has the advantage of not being restricted to Gaussian signals.

The starting point is, of course, the spectral representation theorem and, in particular, formula (9.4.1), which writes the signal $X(n)$ as a random Fourier coefficient,

$$
X(n) = \int_0^1 e^{j2\pi nf} \, dW(f), \qquad n = 1, 2, \ldots, N, \tag{9.5.2}
$$

of a process $W(f)$ with uncorrelated increments and cumulative control function $C_W(f)$ equal to the desired cumulative spectrum $S_X(f)$.

We will assume that the spectrum of $X(n)$ is (absolutely) continuous; that is, it has a power spectrum density $S_X(f)$ such that

$$C_W(f) = S_X(f) = \int_0^f S_X(g)\,dg. \qquad (9.5.3)$$

For computational purposes, the random integral (9.5.2) has to be discretized. More precisely, we have to choose an integer K, and partition

$$f_0 = 0, \quad f_1 = \frac{1}{K}, \quad f_2 = \frac{2}{K}, \ldots, f_{K-1} = \frac{K-1}{K}, \quad f_K = 1,$$

of the interval $[0, 1]$, and replace the right-hand side of (9.5.2) by the sums

$$X_K(n) = \sum_{k=1}^{K} e^{j2\pi n f_k} \left(W(f_k) - W(f_{k-1}) \right)$$

$$= \sum_{k=1}^{K} e^{j2\pi n(k/K)} \left(W\left(\frac{k}{K}\right) - W\left(\frac{k-1}{K}\right) \right).$$

The increments

$$W\left(\frac{1}{K}\right) - W\left(\frac{0}{K}\right), \quad W\left(\frac{2}{K}\right) - W\left(\frac{1}{K}\right), \quad \ldots, \quad W\left(\frac{K}{K}\right) - W\left(\frac{k-1}{K}\right),$$

are zero-mean and uncorrelated and have, respectively, variances

$$\sigma_1^2 = \int_0^{1/K} S_X(f)\,df, \quad \sigma_2^2 = \int_{1/K}^{2/K} S_X(f)\,df, \quad \ldots, \sigma_K^2 = \int_{(K-1)/K}^{1} S_X(f)\,df.$$

Hence, the total mean powers of $X(n)$ and $X_K(n)$ match exactly. Thus, the simulation algorithm calls for the following steps:

Step 0. Select a positive integer K determining the accuracy of our simulation.
Step 1. Generate, via a random number generator, a sequence

$$\xi_1, \xi_2, \ldots, \xi_K,$$

of zero-mean, variance 1, uncorrelated random values of an otherwise arbitrary distribution.
Step 2. Calculate the variances

$$\sigma_1^2, \sigma_2^2, \ldots, \sigma_K^2$$

defined above via the desired power spectrum density.

Step 3. Calculate the complex numbers

$$x_n = \sum_{k=1}^{K} e^{j2\pi n(k/K)} \sigma_k \xi_k, \qquad n = 1, 2, \ldots, N.$$

They represent an approximate sample of our desired random signal.

Step 4. Plot the real and imaginary parts of the sequence $x_n, n = 1, 2, \ldots, N$,

$$\text{Re } x_n = \sum_{k=1}^{K} \cos(j2\pi n(k/K)) \sigma_k \xi_k, \quad \text{Re } x_n = \sum_{k=1}^{K} \sin(j2\pi n(k/K)) \sigma_k \xi_k,$$

as functions of the variable n.

Remark 9.5.1. It should be observed that if the power spectrum density is symmetric about the midpoint $f = 1/2$, that is, $S_X(1/2 + f) = S_X(1/2 - f)$, then the autocovariance function is real-valued because

$$\gamma_X(n) = \int_0^1 e^{j2\pi nf} S_X(f) \, df = \int_0^1 \cos(2\pi nf) S_X(f) \, df.$$

We shall illustrate the above algorithm on a concrete example implemented in the symbolic manipulation language *Mathematica*.

Example 9.5.1 (Mathematica *simulation of a complex-valued stationary signal*). The goal is to simulate a discrete-time signal $X(n)$, $n = 1, 2, \ldots, 150$, with the spectral density function $S_X(f) = f(1 - f)$, $0 \le f \le 1$, pictured below.

Step 0. Select a positive integer K determining the accuracy of the simulation.

```
In[1] :=   K=100
Out[1] =  100
```

Step 1. Generate, via a pseudo-random number generator, a sequence

$$\xi_1, \xi_2, \ldots, \xi_K$$

of zero-mean, variance 1, uncorrelated random values of an otherwise arbitrary distribution. Here we start with a sample of 100 pseudo-random numbers with the Gaussian, $N[0, 1]$-distribution; see Fig. 9.5.1.

```
In[2] := xi = Table[Random[NormalDistribution[0, 1]], {100}]

Out[2]= {-0.608542, -0.193407, 0.667423, 0.665791, 0.796963, 1.50578,
-1.38957, -2.00677, 0.710005, 3.05874, 0.351129, 0.274176, -0.57993,
-0.317531, -1.9642, 0.418438, -1.21485, 0.311505, 2.14493, -0.665234,
0.440417, -1.24286, 0.217456, -1.48803, -1.66472, 0.720181, 2.09662,
0.751509, -0.748984, 0.203246, -0.490937, 1.91771, -0.696637,
```

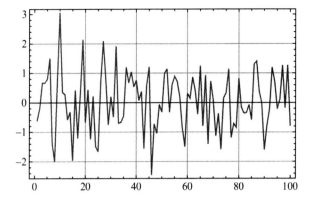

Fig. 9.5.1

```
-0.661528, -0.456505, 1.19835, 0.667494, 1.04284, 0.534665, 0.756436,
0.0707936, 0.375792, -1.56415, 0.559878, 1.20885, -2.45781,
-0.724939, -1.04777, -0.0669847, -0.321047, 0.993232, 1.1395,
-0.325509, 0.611529, 0.890348, 0.716697, 0.203702, -0.863057,
-1.49988, 0.308803, 0.148938, 0.863372, 0.413497, -0.392592, 1.24894,
-0.795932, 0.929254, -1.40817, 0.728825, 0.0811022, -1.13286,
-0.368274, -1.59267, 0.155889, 0.332486, 1.14419, -1.19604,
-0.713426, -0.839724, 0.827024, -0.154212, -0.357799, -0.341499,
-0.0706729, -0.58252, 1.31315, 1.41184, 0.376868, -0.0139196,
-1.60352, -0.783236, -0.223895, 1.19736, 0.707607, -0.212544,
0.115375, 1.27051, -0.18183, 1.27593, -0.775792}

In[2]:= ListPlot[xi, PlotJoined->True, Frame->True,
            GridLines->Automatic]
Out[2]   -Graphics-
```

Step 2. Calculate the standard deviations,

$$\sigma_1, \sigma_2, \ldots, \sigma_K,$$

defined via the above power spectrum density.

```
In[4]:= SX[f_]:=f*(1-f)
In[5]:= sigma =
            Table[ Sqrt[NIntegrate[SX[f], {f, (k-1)/100, (k )/100}]],
            {k,1,100}]

Out[5]= {0.00704746,0.0121518,0.0156098,0.0183757,0.0207284,0.0227962,
    0.0246509,0.0263376,0.0278867,0.0293201,0.030654,0.0319009,0.0330706,
    0.0341711,0.0352089,0.0361893,0.0371169,0.0379956,0.0388287,0.039619,
    0.0403691,0.0410812,0.0417572,0.0423989,0.0430078,0.0435852,0.0441324,
    0.0446505,0.0451405,0.0456034,0.0460398,0.0464507,0.0468366,0.0471982,
    0.047536, 0.0478505,0.0481422,0.0484114,0.0486587,0.0488842,0.0490884,
    0.0492714,0.0494335,0.0495749,0.0496957,0.0497963,0.0498765,0.0499366,
    0.0499767,0.0499967,0.0499967,0.0499767,0.0499366,0.0498765,0.0497963,
```

```
0.0496957,0.0495749,0.0494335,0.0492714,0.0490884,0.0488842,0.0486587,
0.0484114,0.0481422,0.0478505,0.047536,0.0471982,0.0468366,0.0464507,
0.0460398,0.0456034,0.0451405,0.0446505,0.0441324,0.0435852,0.0430078,
0.0423989,0.0417572,0.0410812,0.0403691,0.039619,0.0388287,0.0379956,
0.0371169,0.0361893,0.0352089,0.0341711,0.0330706,0.0319009,0.030654,
0.0293201,0.0278867,0.0263376,0.0246509,0.0227962,0.0207284,0.0183757,
0.0156098,0.0121518,0.00704746}
```

Step 3. Calculate the numbers

$$\text{Re } x_n = \sum_{k=1}^{K} \cos(2\pi n(k/K))\sigma_k \xi_k, \qquad n = 1, 2, \ldots, N,$$

and

$$\text{Im } x_n = \sum_{k=1}^{K} \sin(2\pi n(k/K))\sigma_k \xi_k, \qquad n = 1, 2, \ldots, N,$$

for $N = 150$. They represent approximate samples of the real and imaginary parts of our desired random signal.

```
In[6] ReXi= Table[N[Sum[Cos[2*Pi*n*(k/100)] * sigma[[k]] * xi[[k]],
                          {k,1,100}]],{n,1,150}]

Out[6] = {-0.023415, 0.204973, 0.262053, -0.306833, 0.0423987, 0.0801657,
-0.114673, 0.180827, -0.182326, 0.0501663, 0.241876, -0.422759,
-0.267774, -0.2427, 0.018383, 0.664823, -0.415174, 0.173961,
-0.0833322, 0.197514, -0.078882, 0.203239, 0.00381133, -0.486851,
0.193364, -0.182158, 0.0293311, -0.381732, 0.304001, 0.0549667,
0.410134, -0.0548758, 0.104368, 0.00517703, -0.213219, 0.0621887,
0.122844, 0.119623, -0.21869, -0.00453364, -0.416995, 0.0884643,
0.459038, -0.279907, 0.0401727, -0.216858, 0.00620257, -0.202628,
0.0410997, 0.211609, 0.0410997, -0.202628, 0.00620257, -0.216858,
0.0401727, -0.279907, 0.459038, 0.0884643, -0.416995, -0.00453364,
-0.21869, 0.119623, 0.122844, 0.0621887, -0.213219, 0.00517703,
0.104368, -0.0548758, 0.410134, 0.0549667, 0.304001, -0.381732,
0.0293311, -0.182158, 0.193364, -0.486851, 0.00381133, 0.203239,
-0.078882, 0.197514, -0.0833322, 0.173961, -0.415174, 0.664823,
0.018383, -0.2427, -0.267774, -0.422759, 0.241876, 0.0501663,
-0.182326, 0.180827, -0.114673, 0.0801657, 0.0423987, -0.306833,
0.262053, 0.204973, -0.023415, 0.103954, -0.023415, 0.204973,
0.262053, -0.306833, 0.0423987, 0.0801657, -0.114673, 0.180827,
-0.182326, 0.0501663, 0.241876, -0.422759, -0.267774, -0.2427,
0.018383, 0.664823, -0.415174, 0.173961, -0.0833322, 0.197514,
-0.078882, 0.203239, 0.00381133, -0.486851, 0.193364, -0.182158,
0.0293311, -0.381732, 0.304001, 0.0549667, 0.410134, -0.0548758,
0.104368, 0.00517703, -0.213219, 0.0621887, 0.122844, 0.119623,
-0.21869, -0.00453364, -0.416995, 0.0884643, 0.459038, -0.279907,
0.0401727, -0.216858, 0.00620257, -0.202628, 0.0410997, 0.211609}
```

```
In[7]:= ImXi = Table[N[Sum[Sin[2*Pi*n*(k/100)]*sigma[[k]]*xi[[k]]
         , {k, 1, 100}]], {n, 1, 150}]
```

```
Out[7] = {0.15977, -0.103151, -0.157232, 0.333, -0.139511, 0.245695,
-0.43247, 0.407358, -0.70167, -0.0945059, 0.27421, 0.58988, 0.0705348,
-0.11186, -0.0567596, -0.0596612, -0.574812, -0.467159, 0.0811688,
0.38486, -0.463603, 0.178059, 0.791538, -0.0854149, -0.0661586,
-0.106904, 0.0448853, 0.110552, -0.261648, -0.19714, -0.26017,
0.357341, -0.276876, 0.314915, 0.108389, -0.143431, -0.232836,
-0.121447, 0.474415, -0.426709, 0.176697, -0.123609, -0.138301,
0.132275, 0.660073, -0.661418, -0.361657, 0.239999, -0.134132, 0.,
0.134132, -0.239999, 0.361657, 0.661418, -0.660073, -0.132275,
0.138301, 0.123609, -0.176697, 0.426709, -0.474415, 0.121447,
0.232836, 0.143431, -0.108389, -0.314915, 0.276876, -0.357341,
0.26017, 0.19714, 0.261648, -0.110552, -0.0448853, 0.106904,
0.0661586, 0.0854149, -0.791538, -0.178059, 0.463603, -0.38486,
-0.0811688, 0.467159, 0.574812, 0.0596612, 0.0567596, 0.11186,
-0.0705348, -0.58988, -0.27421, 0.0945059, 0.70167, -0.407358,
0.43247, -0.245695, 0.139511, -0.333, 0.157232, 0.103151, -0.15977,
0., 0.15977, -0.103151, -0.157232, 0.333, -0.139511, 0.245695,
-0.43247, 0.407358, -0.70167, -0.0945059, 0.27421, 0.58988,
0.0705348, -0.11186, -0.0567596, -0.0596612, -0.574812, -0.467159,
0.0811688, 0.38486, -0.463603, 0.178059, 0.791538, -0.0854149,
-0.0661586, -0.106904, 0.0448853, 0.110552, -0.261648, -0.19714,
-0.26017, 0.357341, -0.276876, 0.314915, 0.108389, -0.143431,
-0.232836, -0.121447, 0.474415, -0.426709, 0.176697, -0.123609,
-0.138301, 0.132275, 0.660073, -0.661418, -0.361657, 0.239999,
-0.134132, 0.}
```

Step 4. Plot the complex-valued sequence x_n as a function of the variable n. The consecutive values of the real (left plot) and imaginary (right plot) parts of the numbers $x_1, \ldots, x_{150}$ were joined in Fig. 9.5.2 to better show their progression in time.

```
In[8]:= ListPlot[ReXi, PlotJoined->True, Frame->True,
           GridLines->Automatic]
Out[8]   -Graphics-
In[9]:= ListPlot[ImXi, PlotJoined->True, Frame->True,
           GridLines->Automatic]
Out[9]   -Graphics-
```

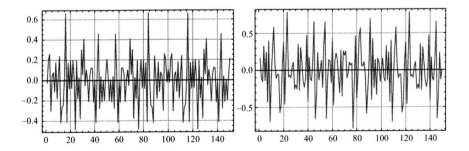

Fig. 9.5.2

Note that for $K = 100$, the smallest frequency present in the representation is $f = 1/100$. Thus, the produced signal sample is periodic with period $P = 100$.

Remark 9.5.2. The above simulation can be adapted to any discrete-time signal $X(t_n)$ with $t_n = n \cdot \Delta t$, extending the procedures described above in the case $\Delta t = 1$ (see Problem 4.5.3). In the theoretical limit, $\Delta t \to 0$, one obtains the spectral representation of continuous time (see Problem 4.5.4).

Remark 9.5.3. The fact that the spectral density was concentrated on the interval $[0, 1]$ was related to the selection of the complex exponentials of the form $e^{j2\pi nf}$ in the spectral representation theorem. A different selection of complex exponentials would lead to different intervals. For example, choosing the complex exponentials of the form $e^{jn\omega}$, that is, conducting spectral analysis in terms of the angular velocity rather than the frequency, would lead to spectral densities concentrated on the interval $[0, 2\pi]$, or any other interval of length 2π. Figure 9.5.3 shows several examples of such spectral densities concentrated on the symmetric frequency interval $[-\pi, +\pi]$, and the real parts of the sample paths of the corresponding stationary signals.

Remark 9.5.4. Another way to produce a graphical representation of the complex-valued signal x_n considered in Example 9.5.1 would be to plot its moduli and arguments instead of its real and imaginary parts.

9.6   Computer Algorithms: Real-Valued Case

To produce a sample of a real-valued stationary discrete-time signal, it is not enough to take a real part of the complex-valued signal, because the real part of a complex-valued stationary signal need not be stationary at all. Indeed, as we have observed before (see Example 4.1.9), the simple complex random harmonic oscillation

$$X(n) = A \cdot e^{j2\pi f_0 n},$$

with the zero-mean random amplitude A, is stationary, with the ACvF

$$EX^*(n)X(n + \tau) = E\left(A^* e^{-j2\pi f_0 n} \cdot A e^{j2\pi f_0(n+\tau)}\right) = E|A|^2 \cdot e^{j2\pi f_0 \tau} = \gamma_X(\tau),$$

but its real part,

$$\mathrm{Re}X(n) = A \cdot \cos(2\pi f_0 n),$$

is not because

$$E[\mathrm{Re}X^*(n) \cdot \mathrm{Re}X(n + \tau)] = E|A|^2 \cdot \cos(2\pi f_0 n)\cos(2\pi f_0(n + \tau))$$
$$= \frac{1}{2}E|A|^2\left(\cos(2\pi f_0(2n + \tau)) + \cos(2\pi f_0 \tau)\right)$$

obviously depends not only on the time lag τ, but also on the time n.

The solution here becomes clear if we abandon the complex domain altogether and return full circle to the very first examples of stationary signals discussed in Chap. 4 (see Examples 4.1.2 and 4.1.3), this time considering them in discrete time, $n = \ldots, -1, 0, 1, \ldots$. Without again going through all the rigorous mathematical details developed in the complex case earlier in this chapter, we just present the basic algorithm.

Consider the real-valued superposition of harmonic oscillations with distinct frequencies $f_1, \ldots, f_K$,

$$X_K(n) = \sum_{k=1}^{K} A_k \cos(2\pi f_k (n + \Theta_k)), \qquad (9.6.1)$$

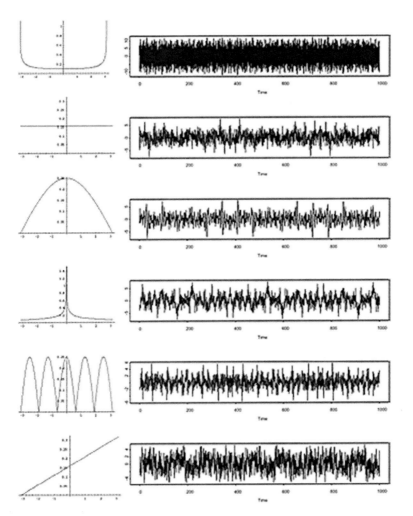

Fig. 9.5.3 Examples of real parts of simulated discrete-time stationary signals (*right column*) with prescribed spectral density functions (*left column*). Note that the spectral densities in these simulations are even and concentrated on the interval $[-\pi, +\pi]$

where $A_k, k = 1, \ldots, K$ are independent, zero-mean real-valued random amplitudes, and $\Theta_k, k = 1, \ldots, K$, are independent random phases, independent of the amplitudes and uniformly distributed over the corresponding periods $P_k = 1/f_k, k = 1, \ldots, K$.

The signal $X_K(n)$ has, obviously, zero mean, and (following calculations analogous to those in Example 4.1.2) the autocovariance sequence is

$$
\begin{aligned}
& \mathbf{E} X_K(n) X_K(n + \tau) \\
&= \mathbf{E} \sum_{k=1}^{K} A_k \cos(2\pi f_k(n + \Theta_k)) \cdot \sum_{l=1}^{K} A_k \cos(2\pi f_l(n + \tau + \Theta_l)) \\
&= \sum_{k=1}^{K} \sum_{l=1}^{K} \mathbf{E}\Big[\Big(A_k \cos(2\pi f_k(n + \Theta_k)) \Big) \Big(A_l \cos(2\pi f_l(n + \tau + \Theta_l)) \Big) \Big] \\
&= \sum_{k=1}^{K} \mathbf{E} A_k^2 \cdot \mathbf{E}\Big(\cos(2\pi f_k(n + \Theta_k)) \cdot \cos(2\pi f_k(n + \tau + \Theta_k)) \Big).
\end{aligned}
$$

Taking into account the trigonometric formula for the product of the cosines in Table 1.3.1, and the uniform distributions of Θ_k over the periods P_k, we finally obtain the autocovariance sequence

$$
\gamma_{X_K}(\tau) = \mathbf{E} X_K(n) X_K(n + \tau) = \frac{1}{2} \sum_{k=1}^{K} \mathbf{E} A_k^2 \cdot \cos(2\pi f_k \tau). \tag{9.6.2}
$$

The above autocovariance sequence corresponds to the power spectrum (see Example 5.2.2)

$$
S_{X_K}(f) = \frac{1}{4} \sum_{k=1}^{K} \mathbf{E} A_k^2 \cdot \Big(\delta(f - f_k) + \delta(f + f_k) \Big). \tag{9.6.3}
$$

Now, let us consider an arbitrary even power spectrum $S_X(f)$, i.e., satisfying condition

$$
S_X(-f) = -S_X(f),
$$

and restricted, for the sake of convenience, to the symmetric interval $[-1/2, +1/2]$. The strategy is to approximate $S_X(f)$ by $S_{X_K}(f)$ described in (9.6.3), while preserving the total power, that is, requiring that

$$
PW_X = \int_{-1/2}^{+1/2} S_X(f) \, df = \int_{-1/2}^{+1/2} S_K(f) \, df = PW_{X_K} = \gamma_{X_K}(0). \tag{9.6.4}
$$

The frequencies f_k will now be taken to correspond to the partition of the interval $[0, 1/2]$, that is,

$$
f_k = \frac{1}{2} \cdot \frac{k}{K}, \quad k = 1, \ldots, K. \tag{9.6.5}
$$

So it suffices to select the random amplitudes A_k, so that

$$\frac{1}{4}\mathbf{E}A_k^2 = \int_{f_{k-1}}^{f_k} S_X(f)\,df \equiv \sigma_k^2, \quad k = 1,\ldots,K. \tag{9.6.6}$$

As a result,

$$PW_{X_K} = \frac{1}{2}\sum_{k=1}^{K}\mathbf{E}A_k^2 = 2\sum_{k=1}^{K}\int_{f_{k-1}}^{f_k} S_X(f)\,df = 2\int_0^{1/2} S_X(f)\,df = PW_X,$$

as required.

Now if we start with realizations of the standard white noise, $\xi_k = W(k), k = 1,\ldots,K$ (of any distribution), and the white noise $\eta_k = U(k), k = 1,\ldots,K$, uniformly distributed on $[-1/2,+1/2]$, then

$$\theta_k = \frac{2K}{k}\eta_k, \quad k = 1,\ldots,K,$$

form a realization of independent random quantities, uniformly distributed over the intervals

$$\left[-\frac{K}{k},+\frac{K}{k}\right] = \left[-\frac{P_k}{2},+\frac{P_k}{2}\right], \quad k = 1,\ldots,K,$$

respectively, and the superposition of real-valued harmonic oscillations,

$$X_K(n) = \sum_{k=1}^{K}\sigma_k\xi_k \cos\left(2\pi f_k(n + 2K\eta_k/k)\right), \tag{9.6.7}$$

will approximate, in the mean-square sense, the signal $X(n)$ with the required power spectrum $S_X(f)$.

Example 9.6.1 (Mathematica simulation of a real-valued stationary signal). We will implement the above algorithm for the power spectrum $S_X(f) = 100f^2$, $-1/2 \le f \le +1/2$, and a Gaussian signal. The general outlines are the same as in Example 9.5.1. First we obtain a sample ξ_k of length $K = 100$ of the standard Gaussian white noise:

```
In[1]:= xi = Table[Random[NormalDistribution[0, 1]], {100}]
```

```
Out[1]= {-0.856053, 1.08187, 2.46229, 0.714797, 0.714182, -0.213566,
-0.433184, -0.851746, -0.0462548, 1.50339, -1.51236, -1.28448,
0.0673793, -0.108364, 0.270925, -0.330244, 1.35095, -0.44158,
-0.357206, -0.647803, -1.09377, -1.34072, 0.849032, 0.0500218,
-0.575234, -0.0171291, -1.79476, 1.31388, -0.628999, -0.593384,
-0.464793, 1.90548, 0.691585, -0.426236, -0.420072, 0.133262,
-0.0273259, -0.499321, -0.169682, -0.91716, 1.63794, 0.746604,
0.0121301, 0.997426, 1.3202, -0.510749, -0.198871, -0.439695,
0.908916, 1.75012, -0.244048, 0.0384926, 0.182402, 0.00244352,
```

```
-2.0007, 0.259864, -0.755299, -1.06697, 0.177168, 0.518347, 0.127846,
-0.426915, 0.831972, 0.130949, -0.708484, 0.744263, 0.0306772,
-2.40272, -0.388865, 1.04692, -2.36268, 1.26858, 0.020974, -1.19099,
-0.0972772, -1.11214, -0.253469, -1.07956, -1.73907, 1.55135,
-0.273338, 0.814078, 0.280743, 0.199324, 1.59616, -0.569614,
-1.32923, -0.0159629, 1.58278, -0.966994, -1.19754, -1.77986,
1.41761, -1.27518, 0.322685, -0.398681, 1.02684, -0.735058,
-0.141971, -0.41919}
```

The next step is to produce a sample of length $K = 100$ of the white noise η_k uniformly distributed on the interval $-1/2, +1/2$. This is accomplished by first producing a white noise uniformly distributed on $[0, 1]$ and then subtracting 1/2 from each of its terms:

```
In[2]:= eta = Table[Random[Real, {0, 1}] - 1/2, {100}]

Out[2]= {0.041948, 0.484289, -0.318925, -0.0276171, 0.0359713,
-0.088659, 0.252302, 0.353539, 0.255555, 0.089573, -0.0901944,
0.227213, 0.0284539, -0.273957, 0.441175, -0.189807, -0.0364003,
0.273394, 0.445258, -0.40948, 0.152135, -0.333722, -0.124852,
-0.42935, -0.389813, -0.318011, -0.305928, 0.0982668, 0.0742158,
0.270648, -0.0582301, 0.244727, 0.318661, -0.318925, -0.468036,
-0.482485, -0.209793, 0.455031, -0.409211, 0.207322, 0.326608,
-0.318363, -0.354468, 0.116801, -0.325528, -0.484641, 0.270384,
0.0461516, -0.435715, 0.33337, 0.0763118, 0.447885, -0.00993046,
-0.437278, -0.365458, -0.296843, 0.171408, 0.381647, -0.397422,
-0.314357, -0.118799, 0.426616, -0.488212, -0.0216788, 0.0545938,
0.244979, 0.366257, 0.36152, -0.119879, 0.22962, -0.404127,
-0.184632, -0.184164, 0.39625, 0.0195609, -0.132516, 0.325766,
0.333528, -0.114981, -0.335674, -0.345642, 0.451881, -0.217559,
0.478683, 0.273157, -0.474735, -0.229347, 0.000362175, -0.281437,
-0.219714, -0.095604, 0.138842, 0.338442, 0.0506663, -0.191477,
-0.176526, 0.0226057, 0.154416, 0.288962, 0.45599}
```

The standard deviations σ_k are

```
In[3]:=sigma=Table[ Sqrt[NIntegrate[100 f^2, {f, (k - 1)/200, (k )/200}]],
                    {k, 1, 100}]
Out[3]={0.00204124, 0.00540062, 0.00889757, 0.0124164, 0.0159426,
0.0194722, 0.0230036, 0.0265361, 0.0300694, 0.0336031, 0.0371371,
0.0406714, 0.044206, 0.0477406, 0.0512754, 0.0548103, 0.0583452, 0.0618803,
0.0654153, 0.0689505, 0.0724856, 0.0760208, 0.0795561, 0.0830913,
0.0866266, 0.0901619, 0.0936972, 0.0972325, 0.100768, 0.104303,
0.107839, 0.111374, 0.114909, 0.118445, 0.12198, 0.125516, 0.129051,
0.132586, 0.136122, 0.139657, 0.143193, 0.146728, 0.150264, 0.153799,
0.157335, 0.16087, 0.164405, 0.167941, 0.171476, 0.175012, 0.178547,
0.182083, 0.185618, 0.189154, 0.192689, 0.196225, 0.19976, 0.203296,
0.206831, 0.210367, 0.213902, 0.217438, 0.220973, 0.224509, 0.228044,
0.23158, 0.235115, 0.238651, 0.242186, 0.245722, 0.249257, 0.252793,
0.256328, 0.259864, 0.263399, 0.266935, 0.27047, 0.274006, 0.277541,
0.281077, 0.284612, 0.288148, 0.291683, 0.295219, 0.298754, 0.30229,
0.305825, 0.309361, 0.312896, 0.316432, 0.319967, 0.323503, 0.327038,
0.330574, 0.33411, 0.337645, 0.341181, 0.344716, 0.348252, 0.351787}
```

Entering the above data in formula (9.6.7) gives us a sample of 150 consecutive values of the desired signal.

```
In[4]:= xn = Table[
        N[Sum[Cos[2*Pi*(n + (200*eta[[k]]/k))*(k/200)]*sigma[[k]]*xi[[k]],
                                        {k, 1, 100}]], {n, 1, 150}]
Out[4] = {0.902888, -1.44987, 0.73034, -0.0467446, -0.981386, 1.85137,
-1.14327, 0.505844, -0.702551, -0.480106, 2.02188, -0.324226,
-2.58959, 3.54, -2.15705, 0.535498, -0.696431, 1.32388, -0.624256,
-1.37706, 0.71946, 0.258689, 2.11401, -2.44596, 1.29879, -1.55369,
0.938206, 0.444133, -0.487131, 0.11673, 0.286159, -1.11297, 0.330835,
-0.246013, -0.0297903, 2.28609, -0.11916, -2.36099, 2.56028,
-2.65337, -0.0894128, 2.2119, -0.816799, 0.344593, -0.698824,
-0.470619, 0.502274, 0.940286, -1.94194, 0.897492, 1.39516, -1.53023,
-0.126247, 0.8947, -0.958154, 0.199293, 0.66053, -1.34534, 1.75322,
-0.338096, -2.11878, 3.2534, -1.21718, -0.405543, 0.413332,
-0.375367, -2.4344, 4.33465, -4.15149, 2.44168, -1.26502, 1.6717,
0.914773, -2.10289, 0.713696, -0.939291, 0.124809, -0.515525,
0.914653, 0.102627, 0.567457, 0.766725, -1.09135, 0.278225, -2.12101,
1.92608, 1.28077, -0.336511, -1.8577, 0.656761, 1.33076, -1.24178,
-0.317488, -0.0655308, 0.540343, 0.00291415, -0.714359, 1.1559,
-1.01383, 0.619388, 1.85065, -3.39279, 2.73494, -1.73749, 0.369481,
0.452425, 0.801605, -1.06127, 1.04946, -1.34485, 0.351694,
-0.0323086, 0.0127435, -1.72899, 0.569055, 1.27245, -1.53539,
2.53497, -1.98056, 1.01728, 0.252221, -0.123346, -0.963119, 1.52522,
-2.59951, 2.70631, -0.853903, 0.17498, -1.08285, -0.805603, 3.50613,
-4.1166, 2.18343, -2.56471, 2.55596, 0.624361, -2.45507, 2.09628,
-0.794994, -0.201666, 0.713224, 0.803646, -1.89323, 1.88523,
-1.57122, 0.958893, -2.30286, 2.02618, 0.408114, -1.40083}
```

This sample path is then plotted in Fig. 9.6.1. To better visualize its progression in time, the discrete plot points are joined. The required code follows on the next page.

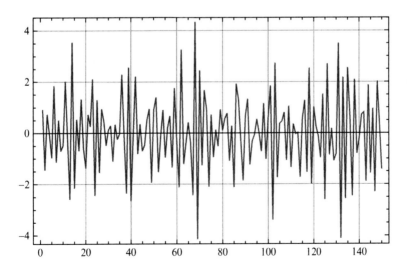

Fig. 9.6.1

```
In[5]:= ListPlot[xn, PlotJoined -> True, Frame -> True,
  GridLines -> Automatic,  PlotStyle -> {Thickness[0.005]}]
```

```
Out[5]=
```

9.7 Problems and Exercises

9.7.1. Verify the polarization formulas preceding the "isometric" formula (9.3.17).

9.7.2. Given a discrete-time stationary signal $X(n)$ with cumulative power spectrum $S_X(f)$, find the cumulative power spectrum for the filtered signal $Y(n) = \sum_{k=-\infty}^{\infty} c_k X(n-k)$. Follow the calculations in Example 9.4.2. Repeat the calculation in the case when $X(n)$ has the power spectral density.

9.7.3. Extend the spectral representation (and the algorithm based on it) in the case of the discrete-time signal $X(t_n)$, with $t_n = n \cdot \Delta t$, extending procedures described above in the case $\Delta t = 1$.

9.7.4. Find the theoretical spectral representation for continuous-time stationary signals taking $\Delta t \to 0$ in Problem 9.7.3.

9.7.5. Use the simulation algorithm described in Sect. 9.5 to produce sample trajectories of complex-valued signals with the following spectral density functions defined on the interval $0 \le f \le 1$. Plot the spectral densities first.

(a)

$$S(f) = \frac{1}{\sqrt{f(1-f)}}$$

What is special about this spectrum? Check that the power of the corresponding signal is finite.

(b)

$$S(f) = 2/3,$$

(c)

$$S(f) = \cos(\pi f),$$

(d)

$$S(f) = 1 - |f|,$$

(e)

$$S(f) = |\sin(8\pi f)|.$$

In the simulations start with (a) the white noise having the $N(0, 1)$ distributions; (b) the white noise having the $U(-1, 1)$ distributions normalized to have variance 1.

9.7.6. Use the simulation algorithm described in Sect. 9.6 to produce sample trajectories of real-valued signals with the following spectral density functions defined on the interval $-1/2 \le f \le +1/2$. Plot the spectral densities first.

(a)

$$S(f) = \frac{1}{\sqrt{(1/2 + f)(1/2 - f)}}$$

(b)

$$S(f) = 2/3$$

(c)

$$S(f) = \cos(\pi f)$$

(d)

$$S(f) = 1 - f$$

(e)

$$S(f) = |\sin(8\pi f)|$$

In the simulations start with (A) the white noise having the $N(0, 1)$ distributions, (B) the white noise having the $U(-1, 1)$ distributions normalized to have variance 1.

9.7.7. Produce plots of several sample paths of the cumulative discrete-time white noise defined in Sect. 9.3. Use (A) the white noise having the $N(0, 1)$ distributions; (B) the white noise having the $U(-1, 1)$ distributions normalized to have variance 1.

9.7.8. Verify that the additivity property (9.3.7) of any continuous function forces its linear form (9.3.8). Start by checking the property for the integers, then move on to rational numbers, and finally extend the result to all real numbers using the continuity assumption.

9.7.9. Figure 9.7.1 shows experimental power spectral densities $S(\lambda)$ of the light emitted by an incandescent lamp at 2,800 degrees Kelvin (on the left), and a fluorescent lamp at 5,000 degrees Kelvin (on the right). The horizontal scale shows the wavelength λ in nanometers.

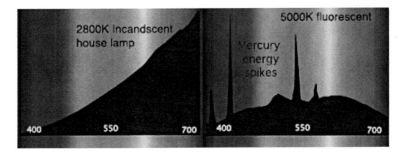

Fig. 9.7.1

(a) Produce approximate mathematical formulas for $S(\lambda)$ representing the above two power spectral densities. Assume an arbitrary vertical scale of the experimental spectra (say, from 0 to 1). Plot them on top of pictures in Fig. 9.7.1 to verify your fit.

(b) Remembering the relationship $f[1/s] \cdot \lambda[m] = c[m/s]$, among the frequency, wavelength, and speed of the travelling wave, and knowing the speed of light, $c[m/s] = 3.0 \cdot 10^8$, convert the approximate mathematical formulas from part (a) to formulas representing the two spectral densities as functions of the frequency f. Plot them.

(c) Use the numerical algorithm from Sect. 9.6 to produce several sample paths of stationary signals with the power spectral densities from (b). Start with the white noise having the $N(0, 1)$ distributions. Plot them.

(d) Do a literature search to comment on whether the selection of the Gaussian distribution in (c) was appropriate for the physical phenomenon under consideration.

Solutions to Selected Problems and Exercises

Chapter 1

Problem 1.5.1. Find the real and imaginary parts of (a) $(j + 3)/(j - 3)$ and (b) $(1 + j\sqrt{2})^3$.

Solution. (a)

$$\frac{j+3}{j-3} = \frac{(j+3)(-j-3)}{(j-3)(-j-3)} = \frac{1-3j-3j-9}{1^2+3^2} = -\frac{4}{5} - \frac{3}{5}j.$$

(b)

$$(1 + j\sqrt{2})^3 = 1^3 + 3 \cdot 1^2(j\sqrt{2}) + 3 \cdot 1(j\sqrt{2})^2 + (j\sqrt{2})^3 = -5 + \sqrt{2}j.$$

Problem 1.5.2. Find the moduli $|z|$ and arguments θ of complex numbers (a) $z = -2j$; (b) $z = 3 + 4j$.

Solution. (a) $|z| = \sqrt{(-2)^2 + 0} = 2$, $\tan\theta = \infty \Rightarrow \theta = -\pi/2$. (You have to be careful with the coordinate angle; here $\cos\theta = 0$, $\sin\theta < 0$.)
(b) $|z| = \sqrt{9 + 16} = 5$, $\tan\theta = 4/3 \Rightarrow \theta = \arctan 4/3$.

Problem 1.5.3. Find the real and imaginary components of complex numbers (a) $z = 5\,e^{j\pi/4}$; (b) $z = -2\,e^{j(8\pi + 1.27)}$.

Solution. (a) $z = 5e^{j\pi/4} = 5\cos(\pi/4) + j\sin(\pi/4) = \frac{5\sqrt{2}}{2} + j\frac{5\sqrt{2}}{2} \Rightarrow \operatorname{Re} z = \frac{5\sqrt{2}}{2}$, $\operatorname{Im} z = \frac{5\sqrt{2}}{2}$.
(b) $z = -2e^{j(8\pi + 1.27)} = -2\cos(1.27) - 2j\sin(1.27) \Rightarrow \operatorname{Re} z = -2\cos(1.27)$, $\operatorname{Im} z = -2\sin(1.27)$.

Problem 1.5.4. Show that

$$\frac{5}{(1-j)(2-j)(3-j)} = \frac{j}{2}.$$

W.A. Woyczyński, *A First Course in Statistics for Signal Analysis*,
DOI 10.1007/978-0-8176-8101-2, © Springer Science+Business Media, LLC 2011

Solution.

$$\frac{5}{(1-j)(2-j)(3-j)} = \frac{5}{(1-3j)(3-j)} = \frac{5}{10j} = \frac{j}{2}.$$

Problem 1.5.5. Sketch sets of points in the complex plane (x, y), $z = x + jy$, such that (a) $|z - 1 + j| = 1$; (b) $z^2 + (z^*)^2 = 2$.

Solution. (a) $\{(x, y) : |z - 1 + j| = 1\} = \{(x, y) : |x + jy - 1 + j| = 1\}$
$= \{(x, y) : |(x - 1) + j(y + 1)| = 1\} = \{(x, y) : (x - 1)^2 + (y + 1)^2 = 1^2\}$.
So the set is a circle with radius 1 and center at $(1, -1)$.
(b) $\{(x, y) : z^2 + (z^*)^2 = 2\} = \{(x, y) : (x + jy)^2 + (x - jy)^2 = 2\} = \{(x, y) : x^2 + 2jxy - y^2 + x^2 - 2jxy - y^2 = 2\} = \{(x, y) : x^2 - y^2 = 1\}$. So the set is a hyperbola (sketch it, please).

Problem 1.5.6. Using de Moivre's formula, find $(-2j)^{1/2}$. Is this complex number uniquely defined?

Solution.

$$(-2j)^{1/2} = \sqrt{2}\left(e^{j(\frac{3\pi}{2} + 2\pi k)}\right)^{1/2} = \sqrt{2}e^{j(\frac{3\pi}{4} + \pi k)}, \qquad k = 0, 1, 2, \ldots$$

$$= \begin{cases} \sqrt{2}e^{j(\frac{3\pi}{4})}, & \text{for } k = 0, 2, 4, \ldots; \\ \sqrt{2}e^{j(\frac{3\pi}{4} + \pi)}, & \text{for } k = 1, 3, 5, \ldots; \end{cases}$$

$$= \begin{cases} \sqrt{2}\left(\cos(\frac{3\pi}{4}) + j\sin(\frac{3\pi}{4})\right), & \text{for } k = 0, 2, 4, \ldots; \\ \sqrt{2}\left(\cos(\frac{7\pi}{4}) + j\sin(\frac{7\pi}{4})\right), & \text{for } k = 1, 3, 5, \ldots. \end{cases}$$

Problem 1.5.10. Using de Moivre's formula, derive the complex exponential representation (1.4.5) of the signal $x(t)$ given by the cosine series representation $x(t) = \sum_{m=1}^{M} c_m \cos(2\pi m f_0 t + \theta_m)$.

Solution.

$$x(t) = c_0 + \sum_{m=1}^{M} c_m \cos(2\pi m f_0 t + \theta_m)$$

$$= c_0 e^{j2\pi 0 f_0 t} + \sum_{m=1}^{M} c_m \frac{1}{2}\left(e^{j(2\pi m f_0 t + \theta_m)} + e^{-j(2\pi m f_0 t + \theta_m)}\right)$$

$$= c_0 e^{j2\pi 0 f_0 t} + \sum_{m=1}^{M} \frac{c_m}{2} e^{j(2\pi m f_0 t + \theta_m)} + \sum_{m=1}^{M} \frac{c_m}{2} e^{-j(2\pi m f_0 t + \theta_m)}$$

$$= \sum_{m=-1}^{-M} \left(\frac{c_{-m}}{2} e^{-j\theta_{-m}}\right) e^{j2\pi m f_0 t} + c_0 e^{j2\pi 0 f_0 t} + \sum_{m=1}^{M} \left(\frac{c_m}{2} e^{j\theta_m}\right) e^{j2\pi m f_0 t}.$$

Problem 1.5.12. Using a computing platform such as *Mathematica*, *Maple*, or *Matlab*, produce plots of the signals

$$x_M(t) = \frac{\pi}{4} + \sum_{m=1}^{M} \left[\frac{(-1)^m - 1}{\pi m^2} \cos mt - \frac{(-1)^m}{m} \sin mt \right],$$

for $M = 0, 1, 2, 3, \ldots, 9$ and $-2\pi < t < 2\pi$. Then produce their plots in the frequency-domain representation. Calculate their power (again, using *Mathematica*, *Maple*, or *Matlab*, if you wish). Produce plots showing how power is distributed over different frequencies for each of them. Write down your observations. What is likely to happen with the plots of these signals as we take more and more terms of the above series, that is, as $M \to \infty$? Is there a limit signal $x_\infty(t) = \lim_{M \to \infty} x_M(t)$? What could it be?

Partial solution. Sample *Mathematica* code for the plot:

```
M = 9;

Plot[
Sum[
    (((-1)^m - 1)/(Pi*m^2))*Cos[m*t]  - (((-1)^m)/m)*Sin[m*t],
        {m, M}],
            {t, -2*Pi, 2*Pi}]
```

Sample power calculation:

```
M2 = 2;

N[Integrate[(1/(2*Pi))*
    Abs[Pi/4 +
        Sum[(((-1)^m - 1)/(Pi*m^2))*Cos[m*u]  - (((-1)^m)/m)*
            Sin[m*u], {m, M2}]]^2, {u, 0, 2*Pi}], 5]

1.4445
```

Problem 1.5.13. Use the analog-to-digital conversion formula (1.1.1) to digitize signals from Problem 1.5.12 for a variety of sampling periods and resolutions. Plot the results.

Solution. We provide sample *Mathematica* code:

```
M=9;
x[t_]:=Sum[(((-1)^m-1)/(Pi*m^2))*Cos[m*t] -
            (((-1)^m)/m)*Sin[m*t], {m,M}]
T=0.1;
R=0.05;

xDigital=Table[R*Floor[x[m T]/R], {m,1,50}];

ListPlot[xDigital]
```

Problem 1.5.14. Use your computing platform to produce a discrete-time signal consisting of a string of random numbers uniformly distributed on the interval [0,1]. For example, in *Mathematica*, the command

```
Table[Random[], {20}]
```

will produce the following string of 20 random numbers between 0 and 1:

```
{0.175245, 0.552172, 0.471142, 0.910891, 0.219577,
0.198173, 0.667358, 0.226071, 0.151935, 0.42048,
0.264864, 0.330096, 0.346093, 0.673217, 0.409135,
0.265374, 0.732021, 0.887106, 0.697428, 0.7723}
```

Use the "random numbers" string as additive noise to produce random versions of the digitized signals from Problem 1.5.12. Follow the example described in Fig. 1.1.3. Experiment with different string lengths and various noise amplitudes. Then center the noise around zero and repeat your experiments.

Solution. We provide sample *Mathematica* code:

```
M=9;
x[t_]:=Sum[(((-1)^m-1)/(Pi*m^2))*Cos[m*t] -
             (((-1)^m)/m)*Sin[m*t], {m,M}]
T=0.1;
R=0.05;
xDigital=Table[R*Floor[x[m T]/R], {m,1,50}];
ListPlot[xDigital]
Noise=Table[Random[], {50}];
noisysig = Table[Noise[[t]] + xDigital[[t]], {t, 1, 50}];
ListPlot[noisysig]
Centernoise = Table[Random[] - 0.5, {50}];
noisysig1 = Table[Centernoise[[t]] + xDigital[[t]], {t, 1,
50}];
ListPlot[noisysig1]
```

Chapter 2

Problem 2.7.1. Prove that the system of real harmonic oscillations

$$\sin(2\pi m f_0 t), \quad \cos(2\pi m f_0 t), \quad m = 1, 2, \ldots,$$

forms an orthogonal system. Is the system normalized? Is the system complete? Use the above information to derive formulas for coefficients in the Fourier expansions in terms of sines and cosines. Model this derivation on calculations in Sect. 2.1.

Solution. First of all, we have to compute the scalar products:

$$\frac{1}{P} \int_0^P \sin(2\pi m f_0 t) \cos(2\pi n f_0 t) dt,$$

$$\frac{1}{P} \int_0^P \sin(2\pi m f_0 t) \sin(2\pi n f_0 t) dt,$$

$$\frac{1}{P} \int_0^P \cos(2\pi m f_0 t) \cos(2\pi n f_0 t) dt.$$

Using the trigonometric formulas listed in Sect. 1.2, we obtain

$$\frac{1}{P} \int_0^P \sin(2\pi m t/P) \cos(2\pi n t/P) \, dt$$

$$= \frac{1}{2P} \int_0^P (\sin(2\pi(m-n)t/P) + \sin(2\pi(m+n)t/P)) \, dt = 0 \text{ for all } m, n;$$

$$\frac{1}{P} \int_0^P \cos(2\pi m t/P) \cos(2\pi n t/P) \, dt$$

$$= \frac{1}{2P} \int_0^P (\cos(2\pi(m-n)t/P) + \cos(2\pi(m+n)t/P)) \, dt$$

$$= \begin{cases} \frac{1}{2}, & \text{if } m = n; \\ 0, & \text{if } m \neq n. \end{cases}$$

Similarly,

$$\frac{1}{P} \int_0^P \sin(2\pi m t/P) \sin(2\pi n t/P) \, dt = \begin{cases} \frac{1}{2}, & m = n; \\ 0, & m \neq n. \end{cases}$$

Therefore, we conclude that the given system is orthogonal but not normalized. It can be normalized by multiplying each sine and cosine by $1/\sqrt{2}$. It is not complete, but it becomes complete, if we add the function identically equal to 1 to it; it is obviously orthogonal to all the sines and cosines.

Using the orthogonality property of the above real trigonometric system, we arrive at the following Fourier expansion for a periodic signal $x(t)$:

$$x(t) = a_0 + \sum_{m=1}^{\infty} [a_m \cos(2\pi m f_0 t) + b_m \sin(2\pi m f_0 t)],$$

with coefficients

$$a_0 = \frac{1}{P} \int_0^P x(t) \, dt,$$

$$a_m = \frac{2}{P} \int_0^P x(t) \cos(2\pi m t/P) \, dt,$$

$$b_m = \frac{2}{P} \int_0^P x(t) \sin(2\pi m t/P) \, dt,$$

for $m = 1, 2, \ldots$.

Problem 2.7.2. Using the results from Problem 2.7.1, find formulas for amplitudes c_m and phases θ_m in the expansion of a periodic signal $x(t)$ in terms of only cosines, $x(t) = \sum_{m=0}^{\infty} c_m \cos(2\pi m f_0 t + \theta_m)$.

Solution. Obviously, $c_0 = a_0$. To find the connection among a_m, b_m and c_m, and θ_m, we have to solve the following system:

$$a_m = c_m \cos\theta_m, \qquad b_m = -c_m \sin\theta_m.$$

This gives us

$$\theta_m = \arctan\left(-\frac{b_m}{a_m}\right), \qquad c_m = \sqrt{a_m^2 + b_m^2}.$$

Problem 2.7.9. Find the complex and real Fourier series for the periodic signal $x(t) = |\sin t|$. Produce graphs comparing the signal $x(t)$ and its finite Fourier sums of order 1, 3, and 6.

Solution. The first observation is that $x(t)$ has period π. So

$$
\begin{aligned}
z_m &= \frac{1}{\pi}\int_0^{\pi} |\sin t| e^{-2jmt}\, dt = \frac{1}{\pi}\int_0^{\pi} \sin t \cdot e^{-2jmt}\, dt \\
&= \frac{1}{\pi}\int_0^{\pi} \frac{e^{jt} - e^{-jt}}{2j}\cdot e^{-2jmt}\, dt = \frac{1}{2j\pi}\int_0^{\pi}\left(e^{jt(1-2m)} - e^{-jt(1+2m)}\right) dt \\
&= \frac{1}{2j\pi}\left(\frac{e^{j\pi(1-2m)} - 1}{j(1-2m)} - \frac{e^{-j\pi(1+2m)}}{-j(1+2m)}\right) dt = \frac{2}{\pi(1-4m^2)},
\end{aligned}
$$

because $e^{j\pi} = e^{-j\pi} = -1$ and $e^{-2jm\pi} = 1$, for all m. Therefore, the sought-after complex Fourier expansion is

$$x(t) = \frac{2}{\pi}\sum_{m=-\infty}^{\infty} \frac{1}{1-4m^2}\cdot e^{j2mt}.$$

We observe that for any $m = \ldots -1, 0, 1, \ldots$, we have $z_{-m} = z_m$. Pairing up complex exponentials with the exponents of opposite signs, and using de Moivre's formula, we arrive at the real Fourier expansion that contains only cosine functions:

$$x(t) = \frac{2}{\pi}\left(1 + 2\sum_{m=1}^{\infty}\frac{\cos(2mt)}{1-4m^2}\right).$$

In particular, the partial sums of orders 1 and 3 are, respectively,

$$s_1(t) = \frac{2}{\pi}\left(1 - \frac{2\cos 2t}{3}\right),$$

$$s_3(t) = \frac{2}{\pi}\left(1 - \frac{2\cos 2t}{3} - \frac{2\cos 4t}{15} - \frac{2\cos 6t}{35}\right).$$

```
x[t_] := Abs[Sin[t]]
pl = Plot[x[t], {t, -2 * Pi, 2 * Pi}, Frame -> True,
    GridLines -> Automatic, PlotStyle -> {Thickness[0.01]}]
```

```
sum[t_, M_] := (2/Pi) * (1 + Sum[(2 / (1 - 4 * m^2)) * Cos[2 * m * t], {m, 1, M}])

s1 = Plot[sum[t, 1],{t, -2 * Pi, 2 * Pi}, Frame -> True,
    GridLines -> Automatic, PlotStyle -> {Thickness[0.01]}]
```

```
s3 = Plot[sum[t, 6],{t, -2 * Pi, 2 * Pi}, Frame -> True,
    GridLines -> Automatic, PlotStyle -> {Thickness[0.01]}]
```

Mathematica code and the output showing $x(t), s_1(t)$, and $s_6(t)$ are shown on the next page.

Problem 2.7.13. (a) The nonperiodic signal $x(t)$ is defined as equal to 1/2 on the interval $[-1, +1]$ and 0 elsewhere. Plot it and calculate its Fourier transform $X(f)$. Plot the latter.

(b) The nonperiodic signal $y(t)$ is defined as equal to $(t + 2)/4$ on the interval $[-2, 0]$, $(-t + 2)/4$ on the interval $[0, 2]$, and 0 elsewhere. Plot it and calculate its Fourier transform $Y(f)$. Plot the latter.

(c) Compare the Fourier transforms $X(f)$ and $Y(f)$. What conclusion do you draw about the relationship of the original signals $x(t)$ and $y(t)$?

Solution. (a) The Fourier transform of $x(t)$ is

$$X(f) = \int_{-1}^{+1} \frac{1}{2} e^{-j2\pi ft}\, dt = \frac{e^{-j2\pi f} - e^{j2\pi f}}{-4j\pi f} = \frac{\sin 2\pi f}{2\pi f}.$$

(b) Integrating by parts, the Fourier transform of $y(t)$ is

$$Y(f) = \int_{-2}^{0} ((t+2)/4)e^{-j2\pi ft}\, dt + \int_{0}^{+2} ((-t+2)/4)e^{-j2\pi ft}\, dt$$

$$= \frac{1}{4}\left(\frac{1}{-j2\pi f} \cdot 2 - \frac{1}{(-j2\pi f)^2}(1 - e^{j2\pi f2}) \right)$$

$$+ \frac{1}{4}\left(\frac{-1}{-j2\pi f} \cdot 2 - \frac{1}{(-j2\pi f)^2}(e^{-j2\pi f2} - 1) \right)$$

$$= \frac{1}{4(-j2\pi f)^2}\left(-(1 - e^{j2\pi f2}) - (e^{-j2\pi f2} - 1) \right)$$

$$= \frac{1}{4(j2\pi f)^2}\left(e^{j2\pi f} - e^{-j2\pi f} \right)^2 = \left(\frac{\sin 2\pi f}{2\pi f} \right)^2.$$

(c) So we have that $Y(f) = X^2(f)$. This means that the signal $y(t)$ is the convolution of the signal $x(t)$ with itself: $y(t) = (x * x)(t)$.

Problem 2.7.18. Utilize the Fourier transform (in the space variable z) to find a solution of the diffusion (heat) partial differential equation

$$\frac{\partial u}{\partial t} = \sigma \frac{\partial^2 u}{\partial z^2},$$

for a function $u(t, z)$ satisfying the initial condition $u(0, z) = \delta(z)$. The solution of the above equation is often used to describe the temporal evolution of the density of a diffusing substance.

Solution. Let us denote the Fourier transform (in z) of $u(t, z)$ by

$$U(t, f) = \int_{-\infty}^{\infty} u(t, z)e^{-j2\pi fz}\, dz.$$

Then, for the second derivative,

$$\frac{\partial^2 u(t, z)}{\partial z^2} \longmapsto (j2\pi f)^2 U(t, f) = -4\pi^2 f^2 U(t, f).$$

So taking the Fourier transform of both sides of the diffusion equation gives the equation

$$\frac{\partial}{\partial t} U(t, f) = -4\pi^2 f^2 \sigma U(t, f),$$

which is now just an ordinary differential linear equation in the variable t, which has the obvious exponential (in t) solution

$$U(t, f) = C e^{-4\pi^2 f^2 \sigma t},$$

where C is a constant to be matched later to the initial condition $u(0, z) = \delta(z)$. Taking the inverse Fourier transform gives

$$u(t, z) = \frac{1}{\sqrt{4\pi\sigma t}} e^{-\frac{z^2}{4\sigma t}}.$$

Indeed, by completing the square,

$$\int_{-\infty}^{\infty} U(t, f) e^{j2\pi fx} \, df = C \int_{-\infty}^{\infty} e^{-4\pi^2 f^2 \sigma t} e^{j2\pi fx} \, df$$

$$= C e^{\frac{-x^2}{4\sigma t}} \int_{-\infty}^{\infty} e^{-4\pi^2 \sigma t (f - jx/(4\pi\sigma))^2} \, df,$$

with the last (Gaussian) integral being equal to $1/\sqrt{4\pi\sigma t}$. A verification of the initial condition gives $C = 1$.

Chapter 3

Problem 3.7.2. Calculate the probability that a random quantity uniformly distributed over the interval $[0, 3]$ takes values between 1 and 3. Do the same calculation for the exponentially distributed random quantity with parameter $\mu = 1.5$, and the Gaussian random quantity with parameters $\mu = 1.5, \sigma^2 = 1$.

Solution. (a) Since X has a uniform distribution on the interval $[0, 3]$, then the value of the p.d.f. is $1/3$ between 0 and 3 and 0 elsewhere.

$$P\{1 \le X \le 3\} = (3 - 1) \cdot 1/3 = 2/3.$$

(b)

$$\int_1^3 \frac{1}{\mu} e^{-x/\mu} dx = \frac{2}{3} \int_1^3 e^{-2x/3} dx = -1(e^{-2\cdot 3/3} - e^{-2/3}) = 0.378.$$

(c) We can solve this problem using the table for the c.d.f. of the standard normal random quantity:

$$P(1 \leq X \leq 3) = P(1 - 1.5 \leq X - \mu \leq 3 - 1.5) = P(-0.5 \leq Z \leq 1.5)$$
$$= \Phi(1.5) - \Phi(-0.5) = 0.9332 - (1 - \Phi(0.5))$$
$$= 0.9332 - 1 + 0.6915 = 0.6247.$$

Problem 3.7.4. The p.d.f. of a random variable X is expressed by the quadratic function $f_X(x) = ax(1 - x)$, for $0 < x < 1$, and is zero outside the unit interval. Find a from the normalization condition and then calculate $F_X(x), EX, \text{Var}(X)$, $\text{Std}(X)$, the nth central moment, and $P(0.4 < X < 0.9)$. Graph $f_X(x)$ and $F_X(x)$.

Solution. (a) We know that for the p.d.f. of any random quantity, we have

$$\int_{-\infty}^{\infty} f_X(x)\, dx = 1.$$

So

$$1 = \int_0^1 ax(1 - x)\, dx = \frac{a}{6}.$$

Thus, the constant $a = 6$.

(b) To find the c.d.f., we will use the definition

$$F_X(x) = \int_{-\infty}^{x} f_X(y)\, dy.$$

In our case, when $0 < x < 1$,

$$F_X(x) = \int_0^x 6y(1 - y)\, dy = x^2(3 - 2x).$$

Finally,

$$F_X(x) = \begin{cases} 0, & \text{for } x < 0; \\ x^2(3 - 2x), & \text{for } 0 \leq x < 1; \\ 1, & \text{for } x \geq 1. \end{cases}$$

(c)

$$EX = \int_0^1 6x^2(1 - x)\, dx = \frac{1}{2},$$

$$\text{Var}(X) = E(X^2) - (EX)^2 = \int_0^1 6x^3(1 - x)\, dx - \frac{1}{4} = \frac{3}{10} - \frac{1}{4} = 0.05,$$

$$\text{Std}(X) = \sqrt{\text{Var}(X)} = \sqrt{0.05} \approx 0.224.$$

(d) The nth central moment is

$$\int_0^1 (x - 0.5)^n 6x(1 - x)\, dx = 6 \int_0^1 x(1 - x) \sum_{k=0}^n x^k \left(-\frac{1}{2}\right)^{n-k} dx$$

$$= 6 \sum_{k=0}^n \binom{n}{k} \left(-\frac{1}{2}\right)^{n-k} \int_0^1 x^{k+1}(1 - x)\, dx$$

$$= 6 \sum_{k=0}^n \binom{n}{k} \left(-\frac{1}{2}\right)^{n-k} \frac{1}{6 + 5k + k^2}.$$

(e)

$$\mathbf{P}(0.4 < X < 0.9) = \int_{0.4}^{0.9} 6x(1 - x)\, dx = 0.62.$$

Problem 3.7.6. Find the c.d.f and p.d.f. of the random quantity $Y = \tan X$, where X is uniformly distributed over the interval $(-\pi/2, \pi/2)$. Find a physical (geometric) interpretation of this result.

Solution. The p.d.f. $f_X(x)$ is equal to $1/\pi$ for $x \in (-\pi/2, \pi/2)$ and 0 elsewhere. So the c.d.f. is

$$F_X(x) = \begin{cases} 0, & \text{for } x \leq -\pi/2; \\ (1/\pi)(x + \pi/2), & \text{for } x \in (-\pi/2, \pi/2); \\ 1, & \text{for } x \geq \pi/2. \end{cases}$$

Hence,

$$F_Y(y) = \mathbf{P}(Y \leq y) = \mathbf{P}(\tan X \leq y) = \mathbf{P}(X \leq \arctan(y))$$

$$= F_X(\arctan(y)) = \frac{1}{\pi}(\arctan(y) + \pi/2).$$

The p.d.f. is

$$f_Y(y) = \frac{d}{dy} F_Y(y) = \frac{d}{dy} \frac{1}{\pi}(\arctan(y) + \pi/2) = \frac{1}{\pi(1 + y^2)}.$$

This p.d.f. is often called the *Cauchy probability density function*.

A physical interpretation: A particle is emitted from the origin of the (x, y)-plane with the uniform distribution of directions in the half-plane $y > 0$. The p.d.f. of the random quantity Y describes the probability distribution of locations of the particles when they hit the vertical screen located at $x = 1$.

Problem 3.7.13. A random quantity X has an even p.d.f. $f_X(x)$ of the triangular shape shown in Sect. 3.7.

(a) How many parameters do you need to describe this p.d.f.? Find an explicit ana-
    lytic formula for the p.d.f. $f_X(x)$ and the c.d.f. $F_X(x)$. Graph both.
(b) Find the expectation and variance of X.
(c) Let $Y = X^3$. Find the p.d.f. $f_Y(y)$ and graph it.

Solution. (a) Notice that the triangle is symmetric about the line $x = 0$. Let us
assume that the vertices of the triangle have the following coordinates: $A(a, 0)$,
$B(-a, 0)$, $C(0, c)$. Then the p.d.f is represented by the equations $y = -\frac{c}{a} x + c$
in the interval $[0, a]$ and $y = \frac{c}{a} x + c$ in the interval $[-a, 0]$. So we need at most
two parameters.

Next, the normalization condition says that area under the p.d.f is 1. So necessarily, $ac = 1 \Rightarrow c = 1/a$. Therefore, actually, one parameter suffices and
our one-parameter family of p.d.f.s has the following analytic description:

$$
f_X(x) = \begin{cases}
0, & \text{for } x < -a; \\
\frac{x}{a^2} + \frac{1}{a}, & \text{for } -a \le x < 0; \\
-\frac{x}{a^2} + \frac{1}{a}, & \text{for } 0 \le x < a; \\
0, & \text{for } x \ge a.
\end{cases}
$$

The corresponding c.d.f. is as follows: If $x < -a$, then $F_X(x) = 0$; if $-a \le$
$x < 0$, then $F_X(x) = \int_{-a}^{x}(\frac{t}{a^2} + \frac{1}{a}) dt = \frac{x^2}{2a^2} + \frac{x}{a} + \frac{1}{2}$; if $0 \le x < a$, then
$F_X(x) = \frac{1}{2} + \int_{0}^{x}(-\frac{t}{a^2} + \frac{1}{a}) dt = \frac{1}{2} - \frac{x^2}{2a^2} + \frac{x}{a}$; if $x > a$, then $F(x) = 1$.
(b) Find the expectation and variance of X.

$$
EX = \int_{-\infty}^{\infty} x\, f_X(x)\, dx = \int_{-a}^{0} x\left(\frac{x}{a^2} + \frac{1}{a}\right) dx + \int_{0}^{a} x\left(-\frac{x}{a^2} + \frac{1}{a}\right) dx = 0.
$$

Of course, the above result can be obtained without any integration by observing
that the p.d.f. is an even function, symmetric about the origin.

$$
\begin{aligned}
\text{Var} X &= \int_{-\infty}^{\infty} x^2\, f_X(x)\, dx = \int_{-a}^{0} x^2\left(\frac{x}{a^2} + \frac{1}{a}\right) dx + \int_{0}^{a} x^2\left(-\frac{x}{a^2} + \frac{1}{a}\right) dx \\
&= \frac{a^2}{6}.
\end{aligned}
$$

(c) The function $y = g(x) = x^3$ is monotone; therefore, there exists an inverse
    function, which in this case is $x = g^{-1}(y) = y^{1/3}$. The derivative $g'(x) = 3x^2$,
    and $g'(g^{-1}(y)) = 3y^{2/3}$. Then [see (3.1.12)]

$$f_Y(y) = \frac{f_X(g^{-1}(y))}{g'(g^{-1}(y))} = \begin{cases} 0, & \text{for } y < (-a)^3; \\ \left(\frac{y^{1/3}}{a^2} + \frac{1}{a}\right)\frac{1}{3y^{2/3}}, & \text{for } (-a)^3 \le y < 0; \\ \left(-\frac{y^{1/3}}{a^2} + \frac{1}{a}\right)\frac{1}{3y^{2/3}}, & \text{for } 0 < y < a^3; \\ 0, & \text{for } y \ge a^3. \end{cases}$$

Here is the needed *Mathematica* code producing the desired plots:

```
(*pdf, a=2*)
H[x_]  := If[x < 0, 0, 1]
f[a_, b_, x_]  := H[x - a] - H[x - b];
ff[x_, a_]  := (x/a^2 + 1/a)*f[-a, 0, x] +
   (-x/a^2 + 1/2)*f[0, a, x]
Plot[ff[x, 2], {x, -3, 3}]

F[x_, a_]  := (x^2/(2*a^2) + x/a + 1/2)*f[-a, 0, x] +
   (1/2 - x^2/(2*a^2) + x/a)*f[0, a, x]
Plot[F[x, 2], {x, -4, 4}]
```

Problem 3.7.15. Verify the Cauchy–Schwartz inequality (3.3.18). *Hint:* Take $Z = (X - \mathbf{E}X)/\sigma(X)$ and $W = (Y - \mathbf{E}Y/\sigma(Y)$, and consider the discriminant of the expression $\mathbf{E}(Z + xW)^2$. The latter is quadratic in the x variable and necessarily always nonnegative, so it can have at most one root.

Solution. The quadratic form in x,

$$0 \le \mathbf{E}(Z + xW)^2 = \mathbf{E}Z^2 + 2x\mathbf{E}(ZW) + x^2\mathbf{E}W^2 = p(x),$$

is nonnegative for any x. Thus, the quadratic equation $p(x) = 0$ has at most one solution (root). Therefore, the discriminant of this equation must be nonpositive, that is,

$$(2\mathbf{E}(ZW))^2 - 4\mathbf{E}W^2\mathbf{E}Z^2 \le 0,$$

which gives the basic form of the Cauchy–Schwarz inequality,

$$|\mathbf{E}(ZW)| \le \sqrt{\mathbf{E}W^2} \cdot \sqrt{\mathbf{E}Z^2}.$$

Finally, substitute for Z and W as indicated in the above hint to obtain the desired result.

Problem 3.7.24. Complete the following sketch of the proof of the central limit theorem from Sect. 3.5. Start with a simplifying observation (based on Problem 3.7.23) that it is sufficient to consider random quantities $X_n, n = 1, 2, \ldots$, with expectations equal to 0, and variances 1.

(a) Define $\mathcal{F}_X(u)$ as the inverse Fourier transform of the distribution of X:

$$\mathcal{F}_X(u) = \mathbf{E}e^{juX} = \int_{-\infty}^{\infty} e^{jux}\, dF_X(x).$$

Find $\mathcal{F}'_X(0)$ and $\mathcal{F}''_X(0)$. In the statistical literature $\mathcal{F}_X(u)$ is called the *characteristic function* of the random quantity X. Essentially, it completely determines the probability distribution of X via the Fourier transform (inverse of the inverse Fourier transform).

(b) Calculate $\mathcal{F}_X(u)$ for the Gaussian $N(0, 1)$ random quantity. Note the fact that its functional shape is the same as that of the $N(0, 1)$ p.d.f. This fact is the crucial reason for the validity of the CLT.

(c) Prove that, for independent random quantities X and Y,

$$\mathcal{F}_{X+Y}(u) = \mathcal{F}_X(u) \cdot \mathcal{F}_Y(u).$$

(d) Utilizing (c), calculate

$$\mathcal{F}_{\sqrt{n}(\bar{X}-\mu_X)/\mathrm{Std}(X)}(u).$$

Then find its limit as $n \to \infty$. Compare it with the characteristic of the Gaussian $N(0, 1)$ random quantity. (Hint: It is easier to work here with the logarithm of the above transform.)

Solution. Indeed, $(X_k - EX_k)/\mathrm{Std}(X_k)$ has expectation 0 and variance 1, so it is enough to consider the problem for such random quantities. Then

(a)

$$\mathcal{F}'_X(0) = \frac{d}{du} Ee^{juX}\bigg|_{u=0} = jEXe^{juX}\bigg|_{u=0} = jEX = 0,$$

$$\mathcal{F}''_X(0) = \frac{d}{du} jEXe^{juX}\bigg|_{u=0} = j^2 EX^2 e^{juX}\bigg|_{u=0} = -EX^2 = -1.$$

(b) If Z is an $N(0, 1)$ random quantity, then

$$\mathcal{F}_Z(u) = \int_{-\infty}^{\infty} e^{juX} \frac{e^{-x^2/2}}{\sqrt{2\pi}} dx = e^{-u^2/2} \int_{-\infty}^{\infty} e^{-\frac{1}{2}(x^2 - 2juX + (ju)^2)} \frac{1}{\sqrt{2\pi}} dx$$

$$= e^{-u^2/2} \int_{-\infty}^{\infty} e^{-\frac{1}{2}(x-ju)^2} \frac{1}{\sqrt{2\pi}} dx$$

$$= e^{-u^2/2} \int_{-\infty}^{\infty} e^{-\frac{1}{2}z^2} \frac{1}{\sqrt{2\pi}} dz = e^{-u^2/2}$$

by changing the variable $x - ju \mapsto z$ in the penultimate integral and because the Gaussian density in the last integral integrates to 1.

(c) Indeed, if X and Y are independent, then

$$\mathcal{F}_{X+Y}(u) = Ee^{ju(X+Y)} = E(e^{juX} \cdot e^{juY}) = Ee^{juX} \cdot Ee^{juY} = \mathcal{F}_X(u) \cdot \mathcal{F}_Y(u)$$

because the expectation of a product of independent random quantities is the product of their expectations.

(d) Observe first that

$$\frac{\sqrt{n}(\bar{X} - \mu_X)}{\text{Std(X)}} = \frac{1}{\sqrt{n}}(Y_1 + \cdots + Y_n),$$

where

$$Y_1 = \frac{X_1 - \mu_X}{\text{Std(X)}}, \ldots, Y_n = \frac{X_n - \mu_X}{\text{Std(X)}},$$

so that, in particular, $Y_1, \ldots, Y_n$ are independent, identically distributed with $EY_1 = 0$ and $EY_1^2 = 1$. Hence, using (a)–(c),

$$\begin{aligned}
\mathcal{F}_{\sqrt{n}(\bar{X} - \mu_X)/\text{Std(X)}}(u) &= \mathcal{F}_{(Y_1/\sqrt{n} + \cdots + Y_n/\sqrt{n})}(u) \\
&= \mathcal{F}_{(Y_1/\sqrt{n})}(u) \cdots \mathcal{F}_{(Y_n/\sqrt{n})}(u) \\
&= [\mathcal{F}_{Y_1}(u/\sqrt{n})]^n.
\end{aligned}$$

Now, for each fixed but arbitrary u, instead of calculating the limit $n \to \infty$ of the above characteristic functions, it will be easier to calculate the limit of their logarithm. Indeed, in view of de l'Hôpital's rule applied twice (differentiating with respect to n; explain why this is okay),

$$\begin{aligned}
& \lim_{n \to \infty} \log \mathcal{F}_{\sqrt{n}(\bar{X} - \mu_X)/\text{Std(X)}}(u) \\
&= \lim_{n \to \infty} \log[\mathcal{F}_{Y_1}(u/\sqrt{n})]^n = \lim_{n \to \infty} \frac{\log \mathcal{F}_{Y_1}(u/\sqrt{n})}{1/n} \\
&= \lim_{n \to \infty} \frac{(1/\mathcal{F}_{Y_1}(u/\sqrt{n})) \cdot \mathcal{F}'_{Y_1}(u/\sqrt{n}) \cdot (-\frac{1}{2}u/n^{3/2})}{-1/n^2} \\
&= \frac{1}{2}u \lim_{n \to \infty} \frac{1 \cdot \mathcal{F}'_{Y_1}(u/\sqrt{n})}{1/n^{1/2}} = \frac{1}{2}u \lim_{n \to \infty} \frac{\mathcal{F}''_{Y_1}(u/\sqrt{n}) \cdot (-\frac{1}{2}u/n^{3/2})}{-\frac{1}{2} \cdot 1/n^{3/2}} \\
&= -\frac{1}{2}u^2,
\end{aligned}$$

because $\mathcal{F}'_{Y_1}(0) = 0$ and $\mathcal{F}''_{Y_1}(0) = -1$; see part (a). So for the characteristic functions themselves,

$$\lim_{n \to \infty} \mathcal{F}_{\sqrt{n}(\bar{X} - \mu_X)/\text{Std(X)}}(u) = e^{-u^2/2},$$

and we recognize the above limit as the characteristic function of the $N(0, 1)$ random quantity; see part (b).

The above proof glosses over the issue of whether indeed the convergence of characteristic functions implies the convergence of c.d.f.s of the corresponding random quantities. The relevant continuity theorem can be found in any of the mathematical probability theory textbooks listed in the Bibliographical Comments at the end of this volume.

Chapter 4

Problem 4.3.1. Consider a random signal

$$X(t) = \sum_{k=0}^{n} A_k \cos\Big(2\pi k f_k(t + \Theta_k)\Big),$$

where $A_0, \Theta_1, \ldots, A_n, \Theta_n$ are independent random variables of finite variance, and $\Theta_1, \ldots, \Theta_n$ are uniformly distributed on the time interval $[0, P = 1/f_0]$. Is this signal stationary? Find its mean and autocovariance functions.

Solution. The mean value of the signal (we use the independence conditions) is

$$\mathbf{E}X(t) = \mathbf{E}\Big(A_1 \cos 2\pi f_0(t + \Theta_1)\Big) + \cdots + \mathbf{E}\Big(A_n \cos 2\pi n f_0(t + \Theta_n)\Big)$$

$$= \mathbf{E}A_1 \cdot \int_0^P \cos 2\pi f_0(t + \theta_1)\frac{d\theta_1}{P} + \cdots + \mathbf{E}A_n \cdot$$

$$\times \int_0^P \cos 2\pi n f_0(t + \theta_n)\frac{d\theta_n}{P} = 0.$$

The mean value doesn't depend on time t; thus, the first requirement of stationarity is satisfied.

The autocorrelation function is

$$\gamma_X(t, t + \tau) = \mathbf{E}[X(t)X(t + \tau)]$$

$$= \mathbf{E}\left(\sum_{i=1}^{n} A_i \cos(2\pi i f_0(t + \Theta_i)) \cdot \sum_{k=1}^{n} A_k \cos(2\pi k f_0(t + \tau + \Theta_k))\right)$$

$$= \sum_{i=1}^{n}\sum_{k=1}^{n}\mathbf{E}(A_i A_k) \cdot \mathbf{E}\Big(\cos(2\pi i f_0(t + \Theta_i)) \cdot \cos(2\pi k f_0(t + \tau + \Theta_k))\Big)$$

$$= \sum_{i=1}^{n} \frac{\mathbf{E}A_i^2}{2} \cos(2\pi i f_0 \tau),$$

because all the cross-terms are zero. The autocorrelation function is thus depending only on τ (and not on t), so that the second condition of stationarity is also satisfied.

Problem 4.3.2. Consider a random signal

$$X(t) = A_1 \cos 2\pi f_0(t + \Theta_0),$$

where A_1, Θ_0 are independent random variables, and Θ_0 is uniformly distributed on the time interval $[0, P/3 = 1/(3f_0)]$. Is this signal stationary? Is the signal $Y(t) = X(t) - \mathbf{E}X(t)$ stationary? Find its mean and autocovariance functions.

Solution. The mean value of the signal is

$$EX(t) = E\left(A\cos 2\pi f_0(t + \Theta)\right) = EA \cdot \int_0^{P/3} \cos(2\pi f_0(t + \theta)) \frac{d\theta}{P/3}$$

$$= \frac{3EA}{2\pi} \sin(2\pi f_0(t+\theta)) \Big|_{\theta=0}^{P/3} = \frac{3EA}{2\pi}\left(\sin(2\pi f_0(t + P/3)) - \sin(2\pi f_0 t)\right).$$

Since

$$\sin p - \sin q = 2\cos\frac{p+q}{2}\sin\frac{p-q}{2},$$

we finally get

$$EX(t) = EA\frac{3\sqrt{3}}{2\pi}\cos\left(2\pi f_0 t + \frac{\pi}{3}\right),$$

which clearly depends on t in an essential way. Thus, the signal is not stationary.

The signal $Y(t) = X(t) - EX(t)$ obviously has mean zero. Its autocovariance function is

$$\gamma_Y(t,s) = E[X(t)X(s)] - EX(t)EX(s)$$

$$= EA^2 \int_0^{P/3} \cos 2\pi f_0(t + \theta)\cos 2\pi f_0(s + \theta)\frac{3}{P} d\theta - EX(t)EX(s),$$

with $EX(t)$ already calculated above. Since $\cos\alpha\cos\beta = [\cos(\alpha + \beta) + \cos(\alpha - \beta)]/2$, the integral in the first term is

$$\cos 2\pi f_0(t - s) + \frac{3}{4\pi}\left(\sin\left(2\pi f_0\left(t + s + \frac{2}{3f_0}\right)\right) - \sin(2\pi f_0(t + s))\right).$$

Now, $\gamma_Y(t,s)$ can be easily calculated. Simplify the expression (and plot the ACF) before you decide the stationarity issue for $Y(t)$.

Problem 4.3.8. Show that if $X_1, X_2, \ldots, X_n$ are independent, exponentially distributed random quantities with identical p.d.f.s $e^{-x}, x \geq 0$, then their sum $Y_n = X_1 + X_2 + \cdots + X_n$ has the p.d.f. $e^{-y} y^{n-1}/(n-1)!$, $y \geq 0$. Use the technique of characteristic functions (Fourier transforms) from Chap. 3. The random quantity Y_n is said to have the gamma probability distribution with parameter n. Thus, the gamma distribution with parameter 1 is just the standard exponential distribution; see Example 4.1.4. Produce plots of gamma p.d.f.s with parameters $n = 2, 5, 20$, and 50. Comment on what you observe as n increases.

Solution. The characteristic function (see Chap. 3) for each of the X_is is

$$\mathcal{F}_X(u) = Ee^{juX} = \int_0^\infty e^{jux}e^{-x}\, dx = \frac{1}{1 - ju}.$$

In view of the independence of the X_is, the characteristic function of Y_n is necessarily the nth power of the common characteristic function of the X_is:

$$\mathcal{F}_{Y_n}(u) = \mathbf{E} e^{ju(X_1 + \cdots + X_n)} = \mathbf{E} e^{juX_1} \cdot \cdots \cdot \mathbf{E} e^{juX_n} = \frac{1}{(1 - ju)^n}.$$

So it suffices to verify that the characteristic function of the p.d.f. $f_n(u) = e^{-y} y^{n-1}/$
$(n-1)!$, $y \geq 0$, is also of the form $(1 - ju)^{-n}$. Indeed, integrating by parts, we obtain

$$\int_0^\infty e^{juy} e^{-y} \frac{y^{n-1}}{(n-1)!} \, dy = \frac{e^{(ju-1)y}}{ju-1} \cdot \frac{y^{n-1}}{(n-1)!} \Big|_{y=0}^\infty$$

$$+ \frac{1}{1-ju} \int_0^\infty e^{(ju-1)y} \frac{y^{n-2}}{(n-2)!} \, dy.$$

The first term on the right side is zero, so that we get the recursive formula

$$\mathcal{F}_{f_n}(u) = \frac{1}{1-ju} \mathcal{F}_{f_{n-1}}(u),$$

which gives the desired result since $\mathcal{F}_{f_1}(u) = \mathcal{F}_X(u) = (1 - ju)^{-1}$.

Chapter 5

Problem 5.4.5. A stationary signal $X(t)$ has the autocovariance function

$$\gamma_X(\tau) = 16e^{-5|\tau|} \cos 20\pi\tau + 8 \cos 10\pi\tau.$$

(a) Find the variance of the signal.
(b) Find the power spectrum density of this signal.
(c) Find the value of the spectral density at zero frequency.

Solution. (a)
$$\sigma^2 = \gamma_X(0) = 16 + 8 = 24.$$

(b) Let us denote the operation of Fourier transform by $\mathcal{F}$. Then writing perhaps a little informally, we have

$$S_X(f) = \int_{-\infty}^\infty \gamma_X(\tau) e^{-j2\pi f\tau} d\tau = (\mathcal{F}\gamma_X)(f)$$

$$= \mathcal{F}\Big(16e^{-5|\tau|} \cdot \cos(20\pi\tau) + 8\cos(10\pi\tau)\Big)(f)$$

$$= 16 \cdot \Big(\mathcal{F}(e^{-5|\tau|}) * \mathcal{F}(\cos(20\pi\tau))\Big)(f) + 8 \cdot \mathcal{F}(\cos(10\pi\tau))(f).$$

But

$$\mathcal{F}(e^{-5|\tau|})(f) = \frac{2 \cdot 5}{5^2 + (2\pi f)^2} = \frac{10}{25 + (2\pi f)^2}$$

and

$$\mathcal{F}(\cos 20\pi \tau)(f) = \frac{\delta(f + 10) + \delta(f - 10)}{2},$$

so that

$$\left(\mathcal{F}(e^{-5|\tau|}) * \mathcal{F}(\cos(20\pi \tau)) \right)(f)$$

$$= \int_{-\infty}^{\infty} \frac{10}{25 + (2\pi f)^2} * \frac{\delta(f - s + 10) + \delta(f - s - 10)}{2} ds$$

$$= 5 \left[\int_{-\infty}^{\infty} \frac{\delta(s - (f + 10))}{25 + (2\pi f)^2} ds + \int_{-\infty}^{\infty} \frac{\delta(s - (f - 10))}{25 + (2\pi f)^2} ds \right]$$

$$= 5 \left[\frac{1}{25 + 4\pi^2(f + 10)^2} + \frac{1}{25 + 4\pi^2(f - 10)^2} \right],$$

because we know that $\int \delta(f - f_0) X(f) \, df = X(f_0)$. Since $\mathcal{F}(\cos 10\pi \tau)(f) = \delta(f + 5)/2 + \delta(f - 5)/2$,

$$S_X(f) = \frac{80}{25 + 4\pi^2(f + 10)^2} + \frac{80}{25 + 4\pi^2(f - 10)^2} + 48(f + 5) + 48(f - 5).$$

Another way to proceed would be to write $e^{-5|\tau|} \cdot \cos(20\pi\tau)$ as $e^{-5\tau} \cdot (e^{j(20\pi\tau)} - e^{-j(20\pi\tau)})/2$, for $\tau > 0$ (similarly for negative τs), and do the integration directly in terms of just exponential functions (but it was more fun to do convolutions with the Diracdelta impulses, wasn't it?).

(c)

$$S_X(0) = \frac{80}{25 + 4\pi^2 100} + \frac{80}{25 + 4\pi^2 100} + 48(5) + 48(-5) = \frac{160}{25 + 400\pi^2}.$$

Problem 5.4.9. Verify the positive-definiteness (see Remark 5.2.1) of autocovariance functions of stationary signals directly from their definition.

Solution. Let N be an arbitrary positive integer, $t_1, \ldots, t_N \in \mathbf{R}$, and $z_1, \ldots, z_N \in \mathbf{C}$. Then, in view of the stationarity of $X(t)$,

$$\sum_{n=1}^{N} \sum_{k=1}^{N} \gamma(t_n - t_k) z_n z_k^* = \sum_{n=1}^{N} \sum_{k=1}^{N} E[X^*(t) X(t + (t_n - t_k))] z_n z_k^*$$

$$= \sum_{n=1}^{N} \sum_{k=1}^{N} E[X^*(t + t_k) X(t + t_n)] z_n z_k^*$$

$$= \mathbf{E} \sum_{n=1}^{N} \sum_{k=1}^{N} (z_k X(t + t_k))^* \cdot (z_n X(t + t_n))$$

$$= \mathbf{E} \left| \sum_{n=1}^{N} z_n X(t + t_n) \right|^2 \geq 0.$$

Chapter 6

Problem 6.4.1. The impulse response function of a linear system is $h(t) = 1 - t$ for $0 \leq t \leq 1$ and 0 elsewhere.

(a) Produce a graph of $h(t)$.

(b) Assume that the input is the standard white noise. Find the autocovariance function of the output.

(c) Find the power transfer function of the system, its equivalent-noise bandwidth, and its half-power bandwidth.

(d) Assume that the input has the autocovariance function $\gamma_X(t) = 3/(1 + 4t^2)$. Find the power spectrum of the output signal.

(e) Assume that the input has the autocovariance function $\gamma_X(t) = \exp(-4|t|)$. Find the power spectrum of the output signal.

(f) Assume that the input has the autocovariance function $\gamma_X(t) = 1 - |t|$ for $|t| < 1$ and 0 elsewhere. Find the power spectrum of the output signal.

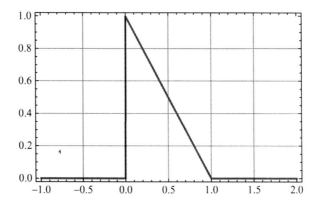

Solution. (a)

(b) With $\gamma_X(\tau) = \delta(\tau)$, the autocovariance function of the output is

$$\gamma_Y(\tau) = \int_0^\infty \int_0^\infty \gamma_X(\tau - u + s)h(s)h(u) \, ds \, du$$

$$= \int_0^1 \int_0^1 \delta(s - (u - \tau))(1 - s)(1 - u) \, ds \, du.$$

As long as $0 < u - \tau < 1$, which implies that $-1 < \tau < 1$, the inner integral is

$$\int_0^1 \delta(s - (u - \tau))(1 - s)\, ds = 1 - (u - \tau),$$

and otherwise it is zero.

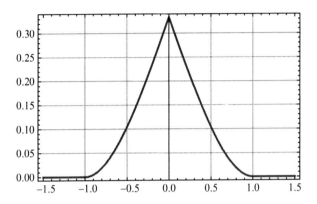

So, for $0 < \tau < 1$,

$$\gamma_Y(\tau) = \int_\tau^1 (1 - (u - \tau))(1 - u)\, du = \frac{1}{6}(\tau - 1)^2(\tau + 2),$$

and, in view of the evenness of the ACvF,

$$\gamma_Y(\tau) = \frac{1}{6}(|\tau| - 1)^2(|\tau| + 2) \quad \text{for } -1 < \tau < 1,$$

and it is zero outside the interval $[-1, 1]$; see the preceding figure.

(c) The transfer function of the system is

$$H(f) = \int_0^1 (1 - t)e^{-2\pi jft}\, dt = \frac{\sin^2(\pi f)}{2\pi^2 f^2} - j\frac{2\pi f - \sin(2\pi f)}{4\pi^2 f^2}.$$

Therefore, the power transfer function is

$$|H(f)|^2 = H(f)H^*(f) = \left(\frac{\sin^2(\pi f)}{2\pi^2 f^2}\right)^2 + \left(\frac{2\pi f - \sin(2\pi f)}{4\pi^2 f^2}\right)^2$$

$$= \frac{-1 + \cos 2\pi f + 2\pi f \sin 2\pi f - 2\pi^2 f^2}{8\pi^4 f^4},$$

as shown in the following figure.

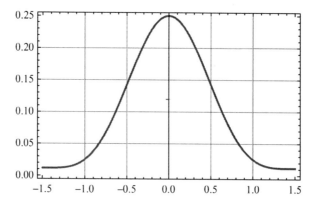

To find the value of the power transfer function at $f = 0$, one can apply l'Hôpital's rule, differentiating the numerator and denominator of $|H(f)|^2$ three times, which yields $|H(0)|^2 = 1/4$. Thus, the equivalent-noise bandwidth is

$$BW_n = \frac{1}{2|H(0)|^2} \int_0^1 (1-t)^2 \, dt = 2/3.$$

Checking the above plot of the power transfer function, one finds that the half-power bandwidth is approximately $BW_{1/2} = 0.553$.

(d) The power spectrum of the output signal is given by

$$S_Y(f) = S_X(f)|H(f)|^2,$$

where $S_X(f)$ is the power spectrum of the input signal. In our case,

$$S_X(f) = \int_{-\infty}^{\infty} \frac{3}{1 + 4t^2} \cdot \cos(2\pi f t) \, dt = \frac{3\pi}{2} e^{-\pi|f|}.$$

Therefore,

$$S_Y(f) = \frac{3\pi}{2} e^{-\pi|f|} \cdot \frac{-1 + \cos 2\pi f + 2\pi f \sin 2\pi f - 2\pi^2 f^2}{8\pi^4 f^4}.$$

(e) In this case, similarly,

$$S_X(f) = \int_{-\infty}^{\infty} e^{-4|t|} \cdot \cos(2\pi f t) \, dt = \frac{2}{4 + \pi^2 f^2}$$

and

$$S_Y(f) = \frac{2}{4 + \pi^2 f^2} \cdot \frac{-1 + \cos 2\pi f + 2\pi f \sin 2\pi f - 2\pi^2 f^2}{8\pi^4 f^4}.$$

(f) Finally, here

$$Sx(f) = \frac{(\sin \pi f)^2}{\pi^2 f^2}$$

and

$$S_Y(f) = \frac{(\sin \pi f)^2}{\pi^2 f^2} \cdot \frac{-1 + \cos 2\pi f + 2\pi f \sin 2\pi f - 2\pi^2 f^2}{8\pi^4 f^4}.$$

Problem 6.4.5. Consider the circuit shown in Fig. 6.4.2. Assume that the input, $X(t)$, is the standard white noise.

(a) Find the power spectra $S_Y(f)$ and $S_Z(f)$ of the outputs $Y(t)$ and $Z(t)$.
(b) Find the cross-covariance,

$$\gamma_{YZ}(\tau) = \mathbf{E}\Big(Z(t)Y(t+\tau)\Big),$$

between those two outputs.

Solution. (a) Note that $X(t) = Y(t) + Z(t)$. The impulse response function for the "Z" circuit is

$$h_Z(t) = \frac{1}{RC} e^{-t/RC},$$

and

$$Y(t) = X(t) - \int_0^\infty h_Z(s)X(t-s)\, ds.$$

So the impulse response function for the "Y" circuit is

$$h_Y(t) = \delta(t) - \int_0^\infty \frac{1}{RC} e^{-s/RC} \delta(t-s)\, ds$$

$$= \delta(t) - \frac{1}{RC} e^{-t/RC}, \qquad t \geq 0.$$

The Fourier transform of $h_Y(t)$ will give us the transfer function

$$H_Y(f) = \int_0^\infty \Big(\delta(t) - \frac{1}{RC} e^{-t/RC}\Big) e^{-2\pi jft}\, dt = \frac{2\pi jRCf}{1 + 2\pi jRCf}.$$

For the standard white noise input $X(t)$, the power spectrum of the output is equal to the power transfer function of the system. Indeed,

$$S_Y(f) = 1 \cdot |H_Y(f)|^2 = \frac{4\pi^2 R^2 C^2 f^2}{1 + 4\pi^2 R^2 C^2 f^2}.$$

The calculation of $S_X(f)$ has been done before, as the "Z" circuit represents the standard RC-filter.

(b)

$$\gamma_{yz}(\tau) = E(Y(t)Z(t+\tau))$$

$$= E\left[\int_{-\infty}^{\infty} X(t-s)h_Y(s)\,ds \int_{-\infty}^{\infty} X(t+\tau-u)h_Z(u)\,du\right]$$

$$= \int_{-\infty}^{\infty}\int_{-\infty}^{\infty} EX(t-s)X(t+\tau-u)h_Y(s)h_Z(u)\,ds\,du$$

$$= \int_0^{\infty}\int_0^{\infty} \delta(\tau-u+s)\left(\delta(s)-\frac{1}{RC}e^{-s/RC}\right)\frac{1}{RC}e^{-u/RC}\,du\,ds$$

$$= \int_0^{\infty}\left(\delta(s)-\frac{1}{RC}e^{-s/RC}\right)\frac{1}{RC}e^{-(\tau+s)/RC}\,ds$$

$$= \int_0^{\infty}\delta(s)\frac{1}{RC}e^{-(\tau+s)/RC}\,ds - \int_0^{\infty}\frac{1}{RC}e^{-s/RC}\frac{1}{RC}e^{-(\tau+s)/RC}\,ds$$

$$= \frac{1}{RC}e^{-\tau/RC} - \frac{1}{2RC}e^{-\tau/RC} = \frac{1}{2RC}e^{-\tau/RC}.$$

Chapter 7

Problem 7.4.2. A signal of the form $x(t) = 5e^{-(t+2)}u(t)$ is to be detected in the presence of white noise with a flat power spectrum of $0.25\ \mathrm{V}^2/\mathrm{Hz}$ using a matched filter.

(a) For $t_0 = 2$, find the value of the impulse response of the matched filter at $t = 0, 2, 4$.
(b) Find the maximum output signal-to-noise ratio that can be achieved if $t_0 = \infty$.
(c) Find the detection time t_0 that should be used to achieve an output signal-to-noise ratio that is equal to 95% of the maximum signal-to-noise ratio discovered in part (b).
(d) The signal $x(t) = 5e^{-(t+2)}u(t)$ is combined with white noise having a power spectrum of $2\ \mathrm{V}^2/\mathrm{Hz}$. Find the value of RC such that the signal-to-noise ratio at the output of the RC filter is maximal at $t = 0.01$ s.

Solution. (a) The impulse response function for the matched filter is of the form

$$h(s) = 5\exp[-(t_0-s+2)]\cdot u(t_0-s) = 5e^{-(4-s)}u(2-s),$$

where t_0 is the detection time and $u(t)$ is the usual unit step function. Therefore,

$$h(0) = 5e^{-4}, \quad h(2) = 5e^{-2}, \quad h(4) = 0.$$

(b) The maximum signal-to-noise ratio at detection time t_0 is

$$\frac{S}{\mathcal{N}}_{\max}(t_0) = \frac{\int_0^\infty x^2(t_0 - s)\, ds}{N_0} = \frac{\int_0^{t_0} 25e^{-2(t_0-s+2)}\, ds}{0.25} = 50e^{-4}(1 - e^{-2t_0}).$$

So

$$\frac{S}{\mathcal{N}}_{\max}(t_0 = 0) = 50e^{-4}.$$

(c) The sought detection time t_0 can thus be found by numerically solving the equation

$$50e^{-4}(1 - e^{-2t_0}) = 0.95 \cdot 50e^{-4},$$

which yields, approximately, $t_0 = -\log 0.05/2 \approx 1.5$.

Chapter 8

Problem 8.5.1. A zero-mean Gaussian random signal has the autocovariance function of the form

$$\gamma_X(\tau) = e^{-0.1|\tau|} \cos 2\pi\tau.$$

Plot it. Find the power spectrum $S_X(f)$. Write the covariance matrix for the signal sampled at four time instants separated by 0.5 s. Find its inverse (numerically; use any of the familiar computing platforms, such as *Mathematica, Matlab*, etc.).

Solution. We will use *Mathematica* to produce plots and do symbolic calculations although it is fairly easy to calculate $S_X(f)$ by direct integration. The plot of $\gamma_X(\tau)$ follows.

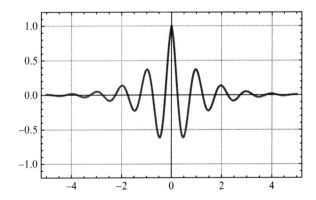

The power spectrum $S_X(f)$ is the Fourier transform of the ACvF, so

```
In[1]:= GX[t_] := Exp[- Abs[t]]*Cos[2*Pi*t];

In[2]:= FourierTransform [GX[t], t, 2*Pi*f]
Out[2]=
```

$$\frac{1}{\sqrt{2\pi}\ (1+4\,(-1+f)^2\,\pi^2)} + \frac{1}{\sqrt{2\pi}\ (1+4\,(1+f)^2\,\pi^2)}$$

Note that the Fourier transform in *Mathematica* is defined as a function of the angular velocity variable $\omega = 2\pi f$; hence the above substitution. The plot of the power spectrum is next.

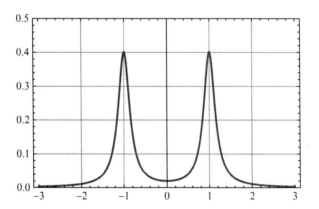

Problem 8.5.3. Find the joint p.d.f. of the signal from Problem 8.5.1 at $t_1 = 1, t_2 = 1.5, t_3 = 2$, and $t_4 = 2.5$. Write the integral formula for

$$P(-2 \le X(1) \le 2, -1 \le X(1.5) \le 4, -1 \le X(2) \le 1, 0 \le X(2.5) \le 3).$$

Evaluate the above probability numerically.

Solution. Again, we use *Mathematica* to carry out all the numerical calculations. First, we calculate the relevant covariance matrix.

```
In[3]:= CovGX = N[{{GX[0], GX[0.5], GX[1], GX[1.5]},
       {GX[0.5], GX[0], GX[0.5], GX[1]},
       {GX[1], GX[0.5], GX[0], GX[0.5]},
       {GX[1.5], GX[1], GX[0.5], GX[0]}}] // MatrixForm
Out[3]=
```

$$\begin{pmatrix} 1. & -0.606531 & 0.367879 & -0.22313 \\ -0.606531 & 1. & -0.606531 & 0.367879 \\ 0.367879 & -0.606531 & 1. & -0.606531 \\ -0.22313 & 0.367879 & -0.606531 & 1. \end{pmatrix}$$

Its determinant and its inverse are

```
In[4]:=Det [CovGX]
Out[4]= 0.25258
```

```
In[5]:= ICovGX = Inverse[CovGX] // MatrixForm
Out[5]=
```

$$\begin{pmatrix} 1.58198 & 0.959517 & -6.73384 \times 10^{-17} & -1.11022 \times 10^{-16} \\ 0.959517 & 2.16395 & 0.959517 & -2.63452 \times 10^{-16} \\ -1.11022 \times 10^{-16} & 0.959517 & 2.16395 & 0.959517 \\ -5.55112 \times 10^{-17} & -2.22045 \times 10^{-16} & 0.959517 & 1.58198 \end{pmatrix}$$

Thus, the corresponding 4D Gaussian p.d.f. is

```
In[6]:=  f[x1, x2, x3, x4]= (1/((2*Pi)^2*Sqrt[Det[CovGX]])) *
     Exp[-(1/2)*
     Transpose[{{x1},{x2},{x3},{x4}}]. ICovGX. {x1,x2,x3,x4}]
Out[6]=  0.05 * E^( -0.79  x1^2 - 1.08  x2^2 - 0.96 x2 x3 -
1.08  x3^2 +  x1 (-0.96 x2 + 8.92*10^-17 x3 + 8.33*10^-17 x4) +
2.43*10^-16 x2 x4 - 0.96 x3 x4 - 0.79  x4^2
```

Note the quadratic form in four variables, x1, x2, x3, x4, in the exponent. The calculation of the sought probability requires evaluation of the 4D integral,

$$P\left(-2 \leq X(1) \leq 2, -1 \leq X(1.5) \leq 4, -1 \leq X(2) \leq 1, 0 \leq X(2.5) \leq 3\right)$$
$$= \int_{-2}^{2} \int_{-1}^{4} \int_{-1}^{1} \int_{0}^{3} f(x_1, x_2, x_3, x_4) \, dx_1 \, dx_2 \, dx_3 \, dx_4,$$

which can be done only numerically:

```
In[7]:= NIntegrate[ f[x1, x2, x3, x4],
          {x1, -2, 2}, {x2, -1, 4}, {x3, -1, 1}, {x4, 0, 3}]
Out[7]= {0.298126}
```

Problem 8.5.4. Show that if a 2D Gaussian random vector $\vec{Y} = (Y_1, Y_2)$ has uncorrelated components Y_1, Y_2, then those components are statistically independent random quantities.

Solution. Recall the p.d.f. of a general zero-mean 2D Gaussian random vector (Y_1, Y_2) [see (8.2.9)]:

$$f_{\vec{Y}}(y_1, y_2) = \frac{1}{2\pi\sigma_1\sigma_2\sqrt{1-\rho^2}} \cdot \exp\left[-\frac{1}{2(1-\rho^2)}\left(\frac{y_1^2}{\sigma_1^2} - 2\rho\frac{y_1y_2}{\sigma_1\sigma_2} + \frac{y_2^2}{\sigma_2^2}\right)\right].$$

If the two components are uncorrelated, then $\rho = 0$, and the formula takes the following simplified shape:

$$f_{\vec{Y}}(y_1, y_2) = \frac{1}{2\pi\sigma_1\sigma_2} \cdot \exp\left[-\frac{1}{2}\left(\frac{y_1^2}{\sigma_1^2} + \frac{y_2^2}{\sigma_2^2}\right)\right];$$

it factors into the product of the marginal densities of the two components of the random vector $\vec{Y}$:

$$f_{\vec{Y}}(y_1, y_2) = \frac{1}{\sqrt{2\pi}\sigma_1} \exp\left[-\frac{1}{2}\left(\frac{y_1^2}{\sigma_1^2}\right)\right] \cdot \frac{1}{\sqrt{2\pi}\sigma_2} \exp\left[-\frac{1}{2}\left(\frac{y_2^2}{\sigma_2^2}\right)\right],$$
$$= f_{Y_1}(y_1) \cdot f_{Y_2}(y_2),$$

which proves the statistical independence of Y_1 and Y_2.

Chapter 9

Problem 9.7.8. Verify that the additivity property (9.3.7) of any continuous function forces its linear form (9.3.8).

Solution. Our assumption is that a function $C(v)$ satisfies the functional equation

$$C(v + w) = C(v) + C(w) \qquad\qquad (S.9.1)$$

for any real numbers v, w. We will also assume that is it continuous although the proof is also possible (but harder) under a weaker assumption of measurability. Taking $v = 0, w = 0$ gives

$$C(0) = C(0) + C(0) = 2C(0),$$

which implies that $C(0) = 0$. Furthermore, taking $w = -v$, we get

$$C(0) = C(v) + C(-v) = 0,$$

so that $C(v)$ is necessarily an odd function.

Now, iterating (S.9.1) n times, we get that for any real number v,

$$C(nv) = n \cdot C(v);$$

choosing $v = 1/n$, we see that $C(1) = nC(1/n)$ for any positive integer n. Replacing n by m in the last equality and combining it with the preceding equality with $v = 1/m$, we get that for any positive integers n, m,

$$C\left(\frac{n}{m}\right) = \frac{n}{m} \cdot C(1).$$

Finally, since any real number can be approximated by the rational numbers of the form n/m, and since C was assumed to be continuous, we get that for any real number,

$$C(v) = v \cdot C(1);$$

that is, $C(v)$ is necessarily a linear function.

Bibliographical Comments

The classic modern treatise on the theory of Fourier series and integrals which influenced much of the harmonic analysis research in the second half of the twentieth century is

[1] A. Zygmund, *Trigonometric Series,* Cambridge University Press, Cambridge, UK, 1959.

More modest in scope, but perhaps also more usable for the intended reader of this text, are

[2] H. Dym and H. McKean, *Fourier Series and Integrals*, Academic Press, New York, 1972,

[3] T. W. Körner, *Fourier Analysis*, Cambridge University Press, Cambridge, UK, 1988,

[4] E. M. Stein and R. Shakarchi, *Fourier Analysis: An Introduction*, Princeton University Press, Princeton, NJ, 2003,

[5] P. P. G. Dyke, *An Introduction to Laplace Transforms and Fourier Series*, Springer-Verlag, New York, 1991.

The above four books are now available in paperback.

The Schwartz distributions (generalized functions), such as the Dirac delta impulse and its derivatives, with special emphasis on their applications in engineering and the physical sciences, are explained in

[6] F. Constantinescu, *Distributions and Their Applications in Physics*, Pergamon Press, Oxford, UK, 1980,

[7] T. Schucker, *Distributions, Fourier Transforms and Some of Their Applications to Physics*, World Scientific, Singapore, 1991,

[8] A. I. Saichev and W. A. Woyczyński, *Distributions in the Physical and Engineering Sciences, Vol. 1: Distributional and Fractal Calculus, Integral Transforms and Wavelets*, Birkhäuser Boston, Cambridge, MA, 1997,

[9] A. I. Saichev and W. A. Woyczyński, *Distributions in the Physical and Engineering Sciences, Vol. 2: Linear, Nonlinear, Fractal and Random Dynamics in Continuous Media*, Birkhäuser Boston, Cambridge, MA, 2005.

Good elementary introductions to probability theory, and accessible reads for the engineering and physical sciences audience, are

[10] J. Pitman, *Probability*, Springer-Verlag, New York, 1993,

[11] S. M. Ross, *Introduction to Probability Models*, Academic Press, Burlington, MA, 2003.

On the other hand,

[12] M. Denker and W. A. Woyczyński, *Introductory Statistics and Random Phenomena: Uncertainty, Complexity, and Chaotic Behavior in Engineering and Science,* Birkhäuser Boston, Cambridge, MA, 1998,

deals with a broader issue of how randomness appears in diverse models of natural phenomena and with the fundamental question of the meaning of randomness itself.

More ambitious, mathematically rigorous treatments of probability theory, based on measure theory, can be found in

[13] P. Billingsley, *Probability and Measure*, Wiley, New York, 1983,

[14] O. Kallenberg, *Foundations of Modern Probability*, Springer-Verlag, New York, 1997,

[15] M. Loève, *Probability Theory*, Van Nostrand, Princeton, NJ, 1961.

All three also contain a substantial account of the theory of stochastic processes.

Readers more interested in the general issues of statistical inference and, in particular, parametric estimation, should consult

[16] G. Casella and R. L. Berger, *Statistical Inference*, Duxbury, Pacific Grove, CA, 2002,

or

[17] D. C. Montgomery and G. C. Runger, *Applied Statistics and Probability for Engineers*, Wiley, New York, 1994.

The classic texts on the general theory of stationary processes (signals) are

[18] H. Cramer and M. R. Leadbetter, *Stationary and Related Stochastic Processes: Sample Function Properties and Their Applications*, Dover Books, New York, 2004,

[19] A. M. Yaglom, *Correlation Thoery of Stationary and Related Random Functions*, Vols. I and II, Springer-Verlag, New York, 1987.

However, the original,

[20] N. Wiener, *Extrapolation, Interpolation, and Smoothing of Stationary Time Series*, MIT Press and Wiley, New York, 1950.

still reads very well.

Statistical tools in the spectral analysis of stationary discrete-time random signals (also known as *time series*) are explored in

[21] P. Bloomfield, *Fourier Analysis of Time Series: An Introduction*, Wiley, New York, 1976,

[22] P. J. Brockwell and R. A. Davis, *Time Series: Theory and Methods*, Springer-Verlag, New York, 1991.

and difficult issues in the analysis of nonlinear and nonstationary random signals are tackled in

[23] M. B. Priestley, *Non-linear and Non-stationary Time Series Analysis,* Academic Press, London, 1988,

[24] W. J. Fitzgerald, R. L. Smith, A. T. Walden, and P. C. Young, eds., *Nonlinear and Nonstationary Signal Processing*, Cambridge University Press, Cambridge, UK, 2000.

The latter is a collection of articles, by different authors, on the current research issues in the area.

A more engineering approach to random signal analysis can be found in a large number of sources, including

[25] A. Papoulis, *Signal Analysis*, McGraw-Hill, New York, 1977,

[26] R. G. Brown and P. Y. Hwang, *Introduction to Random Signal Analysis and Kalman Filtering*, Wiley, New York, 1992.

A general discussion of transmission of signals through linear systems can be found in

[27] M. J. Roberts, *Signals and Systems: Analysis of Signals Through Linear Systems*, McGraw-Hill, New York, 2003,

[28] B. D. O. Anderson, and J. B. Moore, *Optimal Filtering*, Dover Books, New York, 2005.

Gaussian stochastic processes are thoroughly investigated in

[29] I. A. Ibragimov and Y. A. Rozanov, *Gaussian Random Processes*, Springer -Verlag, New York, 1978,

[30] M. A. Lifshits, *Gaussian Random Functions*, Kluwer Academic Publishers, Dordrecht, the Netherlands, 1995.

and for a review of the modern mathematical theory of not necessarily second-order and not necessarily Gaussian stochastic integrals, we refer to

[31] S. Kwapien and W. A. Woyczyński, *Random Series and Stochastic Integrals: Single and Multiple*, Birkhäuser Boston, Cambridge, MA, 1992.

Index